Sculpture in Concrete

by the same author:

Sculpture in ciment fondu (*out of print*)

Sculpture in Concrete

JOHN W MILLS ARCA

London: Maclaren and Sons Ltd

85334 016 1
SCULPTURE IN CONCRETE
BY JOHN W. MILLS
FIRST PUBLISHED 1968
BY MACLAREN AND SONS LTD
7 GRAPE STREET
LONDON WC2

PRINTED BY
THE MILLBROOK PRESS LTD
SOUTHAMPTON

Acknowledgements

I would like to acknowledge all those persons who have contributed in various ways to the production of this book, those from whom I learned certain skills and those who helped with various experiments.

To D. E. L. Parsons of C.R. Books I owe special thanks. It was he who published my first book, back in 1959, and encouraged the writing of this to bring up to date and broaden the scope of that original small work *Sculpture in ciment fondu.*

Sheila Bradbee patiently types all my manuscripts and does the work without fault. Lynton Money has produced photographs of his usual standard and understanding of technique. I could not do without either person.

Mr Geoffrey Robinson has edited this book and seen it through the various complexities and I acknowledge his care and understanding.

I hope this book will provide some help to some sculptors concerned not only with fashion and idiom, but also with skill and technique.

Contents

Illustrations

PLATES

FIGURES IN THE TEXT

Sculpture in Concrete

Introduction

When I wrote *Sculpture in Ciment Fondu* in 1958, it was my belief that the production of a work of sculpture in a hard durable material was a necessity, hardness and durability being the very essence of sculpture. Since then there has been a rapid development of consumer products of an impermanent 'throw away' nature. These have had a strange influence in the visual arts, to the point where some artists make duplicates of household packaged goods. Assemblages of mixed and various materials are commonplace today, in galleries and collections. The director of the Tate Gallery recently remarked that 'he knew; works bought for the nation, were sometimes likely to have a very short life, due to the fragile transient nature of the material; this is a reflection of the times.' In terms of aesthetics this is questionable. It used to be said 'time will tell', but time it seems, may not get the chance.

Despite such developments, I am still of the opinion that to produce a sculpture capable of withstanding the elements, is a desirable achievement. No matter what may be the idiomatic expression or whatever it might reflect, work that proves practical in a public site, in a gallery or collection, having a reasonably long life, contributes to a culture in a positive way.

To produce such work in the studio is, of course, most desirable, ensuring, as it does, direct control by the artist over the production of the image at all times.

Sculptures of various kinds of concrete can be assessed directly in relation to the traditional materials, particularly stone and marble. Hardness, and weight; tooled, cut, abraded, and polished surfaces; the balance of form and the interaction of texture, are some of the qualities most looked for in the technical study of great sculptures of the past.

The imitation of accepted materials by new, has always been a stumbling block, and this practice is not advocated here. There really is no point in making such imitations. The need to do so in the past sprang from a misunderstanding and consequent misuse of new materials. Concrete, because of its extreme variety of application, suffered in this way. Concrete artificial stone is still made for a variety of uses, wall cladding and garden ornaments being two of the worst survivals. In *Sculpture in Ciment Fondu* I described briefly how to make metallic finishes, inadvertantly contributing further misuse. I remember watching one young sculptor trying very hard, and with some success, to make a concrete casting appear like real bronze. It must be admitted however, that it was through such experiments that concrete achieved an accepted place in the repertoire of sculptors' materials.

It is pleasing to see that qualities inherent in the material are now being exploited, such qualities as integral colour, texture, and castability. In the building industry these qualities can be seen in wide variety. Exposed aggregates offer a range of colour and texture not possible with any other material. The fashion for retaining the visual character of the shuttering or formwork in cast concrete columns, walls, ceilings, stairwells etc., is strong evidence of its extreme castability.

Since 1959 there has been a flood of sculpture made of concrete. Every mixed exhibition has a good number of works in these media. Many new buildings have fine examples of cast concrete decoration. Concrete playground sculptures and adventures into three dimensional expression by children using concrete are common today.

In art schools concrete has become one of the most common casting materials. It offers a slower working pace than almost any other casting material used in the sculpture studio. Being able to wash out a faulty filling before it sets, and to make a fresh start, is an advantage unique to concrete.

Sculptural form

A point that must be cleared up, is the use of the term 'mural'. Articles, reviews and appreciations have often described as 'murals' works that adorn buildings, decorating or embellishing facades, by textured pattern, symbols, figures, abstract pattern and form. The correct term is 'relief'. The nature of the work is sculpture. A relief is the projection or recession of a design from a flat surface. The design can be cut back or incised to make an *intaglio* relief, or projected forward to make a *low, half* or *high*-relief. A mural is essentially a coloured flat wall decoration.

Sculpture is three dimensional form, designed in the round and freestanding, or in relief.

The range of form possible in the use of concrete is infinite. It is possible to cast any form that can be modelled, given the necessary factor of casting skill. The image or decoration should however be planned with the nature of the concrete in mind. This should not be an inhibiting factor, but an element that contributes to that proper marriage of idea and material essential to confident work and development. Coarse concrete, for instance, is best used to realize large forms, simple in concept. Exposed aggregates are used to advantage on forms that offer the possibility of grinding, polishing and abrading to make the final form and surface – forms that are simple and perhaps sensual in concept. Fine aggregate concretes are most suitable for casting to achieve finite results of reproduction – finger prints can be easily reproduced.

Concrete can be built up directly by modelling over a carefully prepared armature and building up to the final forms. In this way an amalgam of modelled and carved forms can result, seeming to break down the time-worn dichotomy of thought that separates modelling from carving. Modelled forms, achieved by addition and building are distinctly different from carved forms achieved by the subtraction process of cutting away waste material. Working directly in concrete permits the use of cast, modelled, and carved forms, in fact an amalgam of the full range of historic sculpture techniques.

Cements

The term concrete has become the generic heading under which are grouped any hard insoluble materials made by mixing aggregate with cement as a binding agent. Cements therefore are of the utmost importance. After all, what you buy from the suppliers is cement and aggregate, not concrete (unless, of course, what is known as 'ready mix' is required). So it is essential to know what cement is.

There are basically only two kinds of cement, Portland cement and high alumina cement. The former gained its name because the concrete made with it looked like Portland stone, a typical British stone. These cements are made by heating the raw material, calcium aluminium silicate in the case of Portland, alumina in that of high alumina cement. The raw materials are heated to clinker, which is then finely ground to a powder, to form the cements as we see them in the bag. Rotary cylindrical kilns and grinders are used in this process.

When water is added to cement a chemical action takes place (hydration). The paste or thick cream made by adding water, gradually solidifies. This action of solidifying, changing from a fairly fluid consistency to give a fine dense solid is the character that makes cement such an admirable casting medium. Extreme castability is its most commendable quality for the sculptor. But the solid made by simply adding water to cement has very little strength or resistance to weather until it is mixed with an aggregate to become 'concrete'. A concrete is as strong as its aggregate.

In the building industry a mixture of cement and sand (aggregate) is usually regarded as a 'mortar'. Lime is sometimes added to a Portland mixture to give it greater plasticity and a measure of flexibility. 'Concrete' demands a mixture of cement and sand plus a larger aggregate. Mortar has little compressive strength by comparison with concrete, and is used simply to lay bricks and render wall surfaces. It is largely this factor of compressive strength that concerns the architect and engineer. Tensile strength is given to concrete forms and members by the addition of steel reinforcing. Reinforced concrete is simply concrete with a mild steel reinforcement placed strategically to avoid tensile weakness. This type of concrete is essentially the kind made and used by sculptors. Greater tensile strength is achieved by making a pre-stressed concrete. In this case the reinforcing, usually of high tensile steel-wire or alloy, is placed under stress or compression before the concrete is placed about it. Pre-stressed concrete forms can therefore be finer since they require less bulk, and it is this quality that has made possible the engineering structures typical of the twentieth century.

The sculptor usually achieves the tensile strength so important to him, by making freestanding forms hollow, working on the principle that, by its very nature, a hollow sphere has a greater strength than a solid sphere, and a tube has greater tensile strength than a solid rod. Compressive strength does not usually concern the sculptor in the same way that it must the engineer.

The structural engineer's job is to work on the known facts of stress and compression, and to apply his specialist knowledge and experience to build the dreams of architects and designers. An example is the building of the Sydney Opera House in Australia, where the engineer has made the architect's vision manifest in concrete and steel.

An artist involved in a work of any great size, should always seek the advice of an experienced engineer, and take advantage of his specialized knowledge.

It will be evident from the references to architects, structural engineers, and the building industry, that properly made concrete is a reliable construction material. This means that, given a proper understanding of the principles and practices involved to make a satisfactory concrete sculpture, items can be made to be placed on buildings, in galleries and collections, with the certain knowledge that they will remain intact for at least as long as the building. If the sculpture is given some extra care and attention (cleaning, for example) and not exposed to hooliganism, then it will enjoy and give pleasure for a considerable period of time.

This book is primarily concerned with the castability of concrete, because this is the main function of the material in the sculptor's studio. Casting skills are therefore of great importance. Although building up a work directly in concrete is a positive technique in comparison with casting, it is infrequently used. The skills necessary to build up a form however, are to some extent embodied in the very processes of casting and knowledge of one directly affects the learning of the other. Both techniques are discussed in the following text.

Comparison

For a long time concrete castings were regarded as great solid lumps, difficult to handle, problematic to make, and liable to litter a site by breaking up and becoming unsightly. This may have been true, seven or eight years ago, but it is no longer true today.

There has been a certain gravitation towards the use of various resins, polyesters, epoxies, acrylics etc., by young sculptors. This is understandable because of the possible range of bright colour, dense, transparent and translucent. Very light weight and fair durability are other factors to be considered in their favour. In certain applications they are undoubtedly the most suitable materials. They also have disadvantages that may not occur in concrete. The accomplished sculptor should of course, have a good working knowledge of most materials in order to be able to choose that most suited to a particular need.

Distinct disadvantages common to all resins, are the degree of critical control of temperature and draught, and the correct mixture of chemicals required to ensure a well resolved solid. Temperature and draughts, if not controlled, may cause the monomer of the resin to evaporate, inhibiting gelation and cure, in other words resulting in a sticky jam-like cast surface. Polyester resin is vulnerable to ultra violet light, and will discolour and eventually break down, losing its properties. There are on the market various additives and equipment to ensure as far as possible against failures of various kinds. Such items however bring the comparative costs of casting almost up to those of cast metals. Acrylic resins are resistant to ultra violet light, but are prohibitive in cost, and as is the case with all resins, they can be easily scratched and damaged.

Concrete, on the other hand, cannot be easily scratched, and, if properly designed and reinforced, has sufficient strength to be as strong if not stronger, than resin castings. It is a fraction of the cost of any resin and infinitely easier to handle and position. In terms of large areas of sculptured relief, both materials offer interesting varieties of colour and texture. Concrete reliefs however, can become an integral part of the fabric of the building. Large areas can be cast *in situ* by the contractor to the same concrete specification etc., as the building itself. This is a distinct advantage as resin reliefs have always to be fixed to a small surface.

Schools can use concrete to introduce pupils to sculpture, and a lot of very interesting work has been done by school children. The various resins do not allow the necessary freedom for this.

It will be appreciated that these two materials, new in the repertoire of sculpture, must be weighed up carefully one against the other, to determine which is more suitable in specific circumstances.

1 Moulding

Waste-mould

Sculpture cast of concrete or any other material requires the production of a mould from which the cast can be made. The mould is the 'negative', sometimes referred to as the 'female'. The casting is the 'positive' or 'male'.

There are many kinds of mould, dependant upon the casting medium, and the nature of the form to be cast. Both the number of castings required and the material of the original form, help to determine the type of mould. It may be simple or complex, and made of plaster of Paris, glass fibre, reinforced polyester resin, polyvinyl chloride, expanded polystyrene, foamed polyurethane, clay, or wood.

To cover all such moulding materials and their techniques would require more space than the scope of this book allows. My book *The Technique of Casting for Sculpture** describes various moulds and moulding skills in some detail, and may prove valuable to those wishing to know more about a wider range of techniques. In general, sculptors use the simple plaster of Paris waste-moulding technique, and this has become the basic skill required by any sculptor interested in cast forms.

The waste-mould, as the name suggests, is made in order to produce a single casting, and becomes waste, by virtue of the fact that it is broken away from the cast positive. It is usually made of plaster of Paris, and taken from a soft clay or synthetic clay original.

The single casting produced from a waste mould is usually the 'master cast', from which other kinds of moulds are made to produce more positives to make up an edition. Sculptors usually make and sell editions of six, for economy reasons. Costs can be spread over the edition, easily permitting more sales than from a unique single casting.

The principle of waste-moulding is to cover the clay original with plaster of Paris, reinforced if necessary. In the thickness of the plaster, openings, referred to as 'caps' or 'traps', are made to permit removal of the original clay plus any armature support, and to allow the placing of the filler or casting medium which will make the positive cast. The caps are designed to give the smallest openings, and the most access to the entire mould surface. The greater part of the original is encased in the remainder of the mould, known as the 'main mould'. The design of the mould, dividing it into 'caps' and 'main mould' is of obvious importance. Generally the front of the form is contained in the 'main mould', the openings being kept as much as possible to the back.

Most casting techniques are refinements of these basic principles, so a good working knowledge of waste-moulding is vital. The following descriptions are offered only as a guide to these techniques. Obviously sculptures require individual attention regarding mould design etc., and no two sculptures present quite the same problems.

*Batsford,

An elementary mould, that can be made to give some experience both of mixing plaster of Paris and placing it, is the mould for a simple relief panel. Made of clay or synthetic clay on a modelling board, the panel should leave wide margins on all sides. If this is not possible, and it is impossible to transfer it on to a larger board, then the margins can be extended with clay.

The mould is made by mixing plaster of Paris 1 :1 water and plaster, and covering the clay panel with the mixture to a thickness of approximately 1in.–1 $\frac{1}{2}$ins. (2·5399–3·8099cm.), according to the area of the relief. Plaster of Paris is mixed equal parts of plaster and water, so as to make a thick cream. This cream gradually solidifies, enabling the mould thickness to be built up. Mild steel reinforcing should be placed to give strength to the mould. When the mould has been made to the required thickness, and the plaster hardened, prise the mould from the board and clay. Do this carefully in order not to break the mould. Water poured between the plaster and clay will help to break the suction between the two.

When the mould is free from the board and the clay relief, clean it and prepare it for filling. See mould preparation and release agents (page 14).

To mould from forms in the round or in high relief, the main mould and caps have first to be designed, and marked on the original, drawing the divisions between caps and main mould on the clay with a knife. Along these lines a wall of brass shim (fencing) is composed, using 1/5000ths brass sheet. The divisions may vary from a simple two-piece mould, to an infinitely complex mould with many caps. Whether simple or complex, the principle remains the same.

The *brass shim* is cut into strips approximately 2in.–3in. (5·0799–7·6199cm.) in length, with a width $\frac{1}{4}$in. (0·6350cm.) wider than the planned thickness of the mould. This $\frac{1}{4}$in. (0·6350cm.) is to allow a firm fixing in the clay. The thickness of the mould can vary according to the size of the job in hand. Properly reinforced with mild steel etc., no mould need be more than 1 $\frac{1}{2}$in. (3·8099cm.). Some sculptors take pride in making thin moulds, because they are easier to chip out, and this may come with practice. To begin with, it is better to make a sensible mould thickness. More accidents occur with thin moulds than with thick, so do not be too eager to make thin moulds.

The cut shim is placed carefully along the drawn line, slightly overlapping each piece. In this way a continuous brass wall is made along the drawn line. It should be only slightly thicker than the actual gauge of metal, because of the overlap. It must be tailored to form an even projection from the clay surface, and the top edge must be trimmed and continuous. No corners or sharp projections should be left, as they may cut fingers. Each piece of brass should be pushed well into the clay to at least $\frac{1}{4}$in. (0·6350cm.). This will resist the weight of plaster when it is placed, if this is not done the plaster will dislodge the shim.

The thickness of seam 'flash' on the positive cast, depends to a great extent on the care with which the shim is placed. The thinner the seam 'flash' the better the volumetric accuracy of the cast.

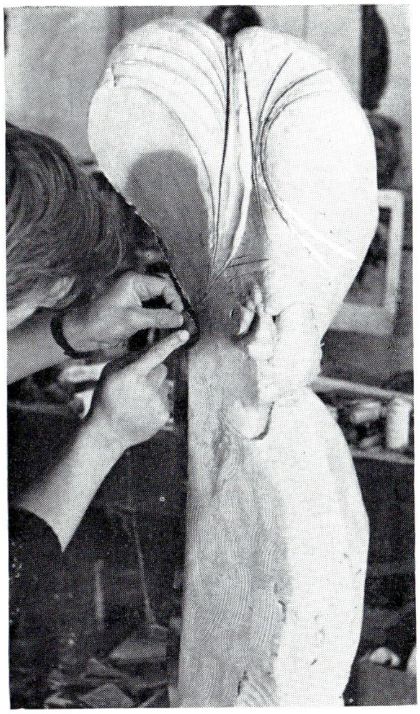

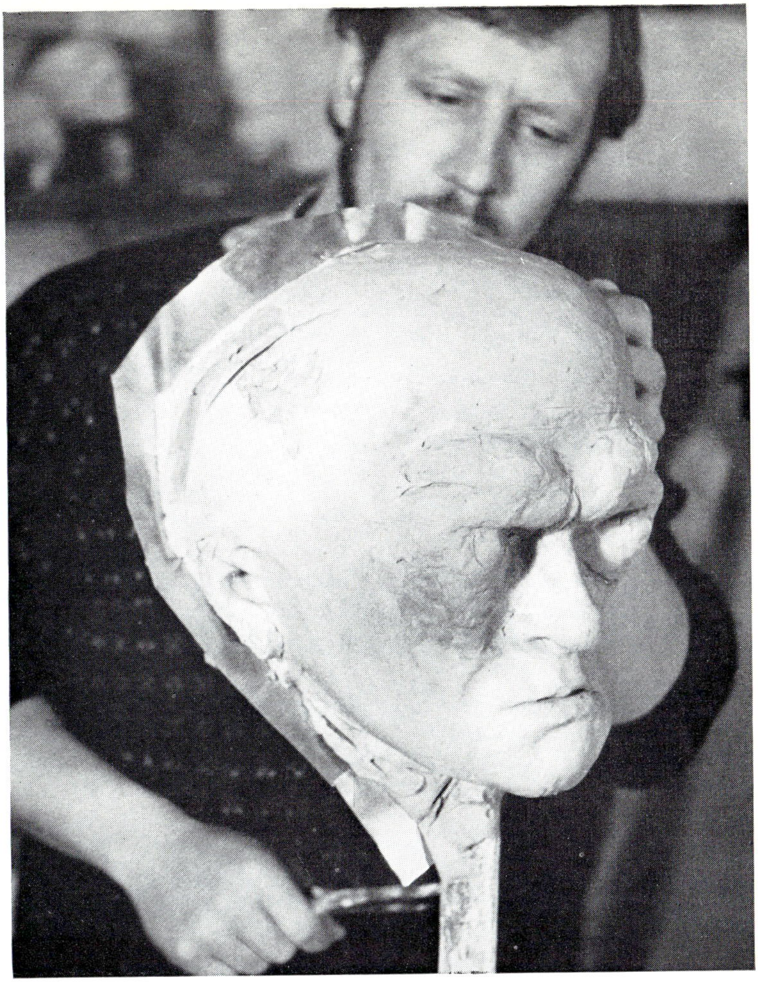

Plates 1, 2 and 3. Brass shim being placed in the clay along the marked line, to make the mould divisions. These need to be placed neatly to ensure a clean division, and to facilitate making an even mould thickness

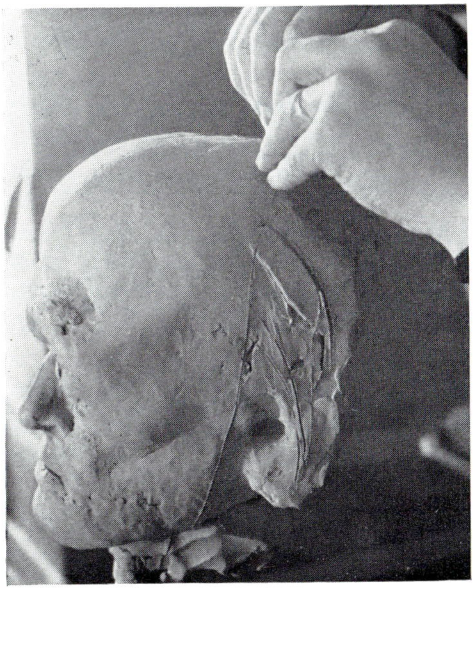

Part of the sequence for making a plaster waste-mould.

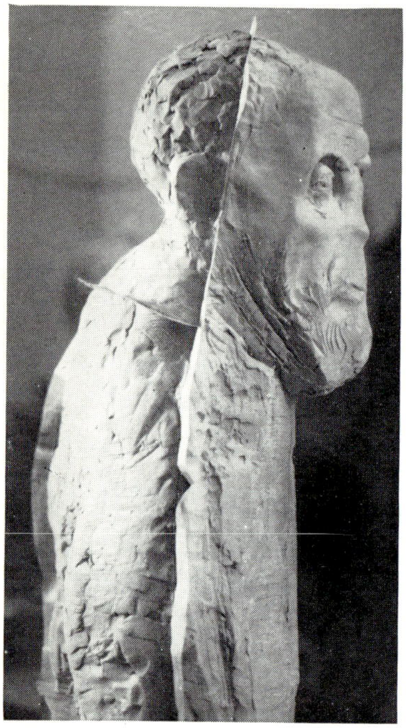

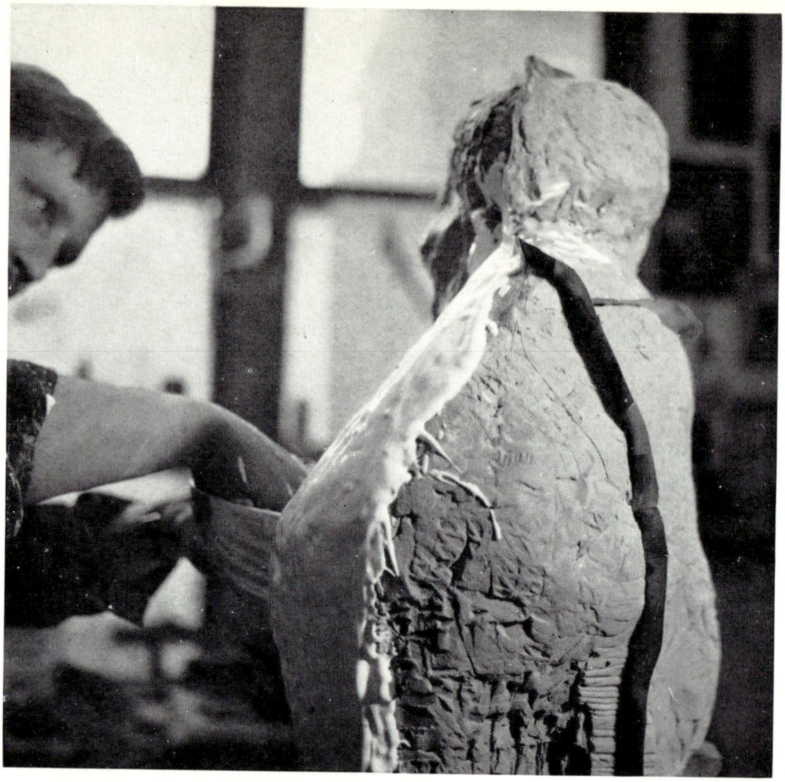

Plate 4. (Above) The brass shim placed to make the mould divisions. Note the registration keys on the vertical seams.

Plate 5. (Top right) Plaster splashed on to the surface to make an even first coat. This could be coloured to make a warning coat.

Plate 6. (Right) The build-up carried on, working out from the seams.

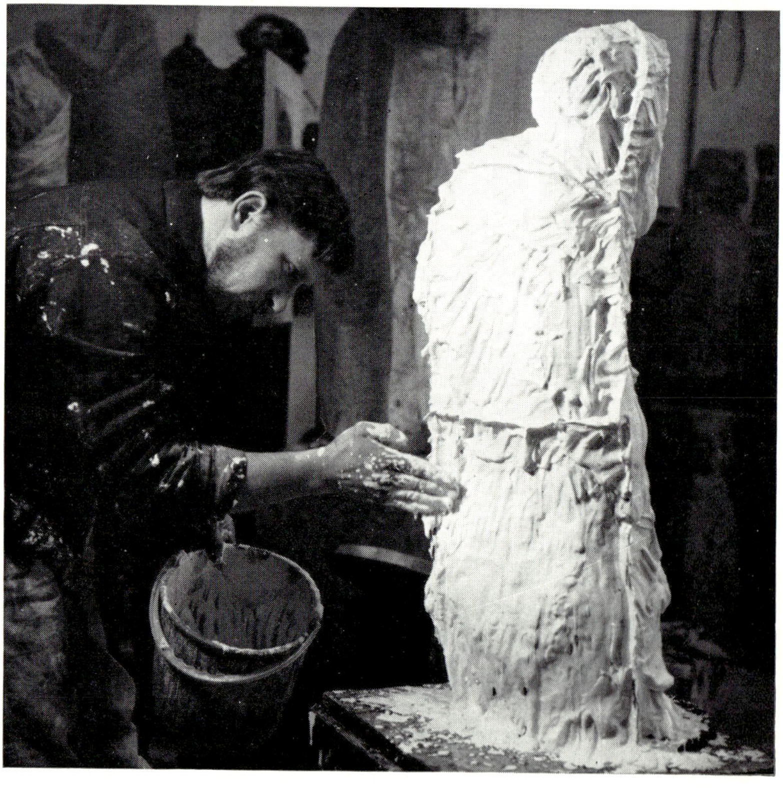

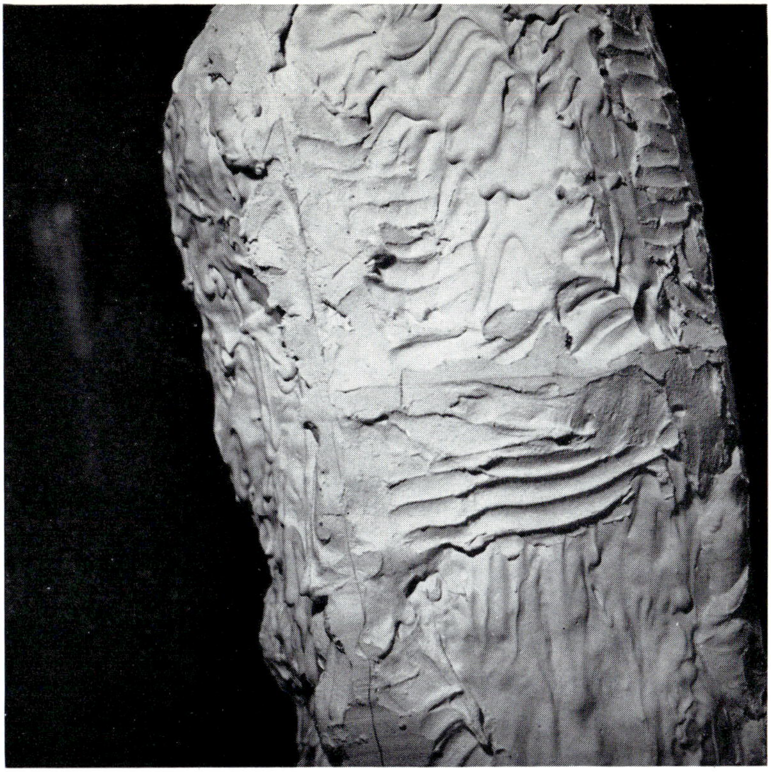

Plate 7. Seams scraped to reveal brass shim after each application of plaster.

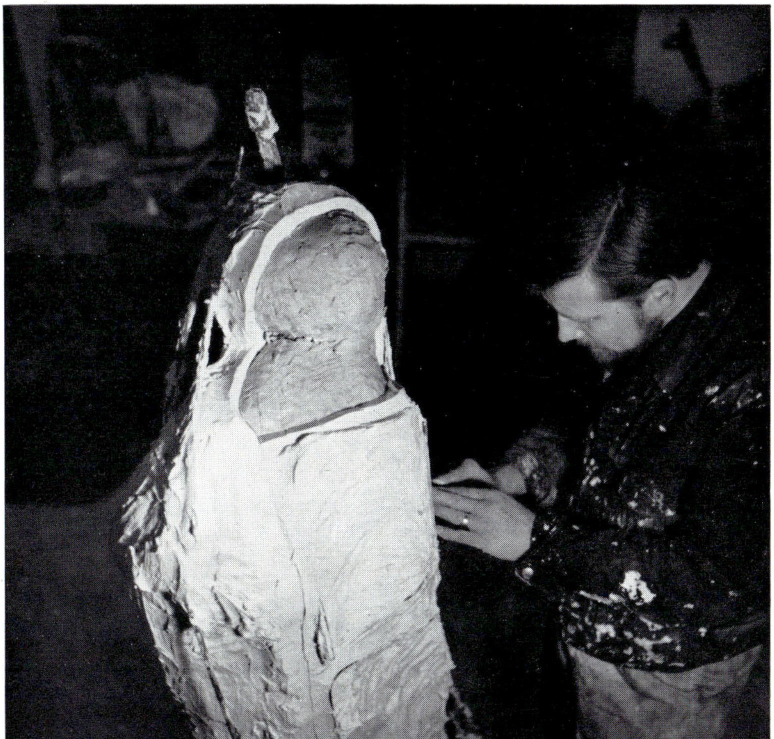

Plate 8. Removing the caps to take out the clay, and place the filling. Metal reinforcing has been included in the plaster build up.

Plate 9. A life size plaster waste-mould ready to be filled. The seam edges can be clearly seen and the cradle with which the mould can be handled.

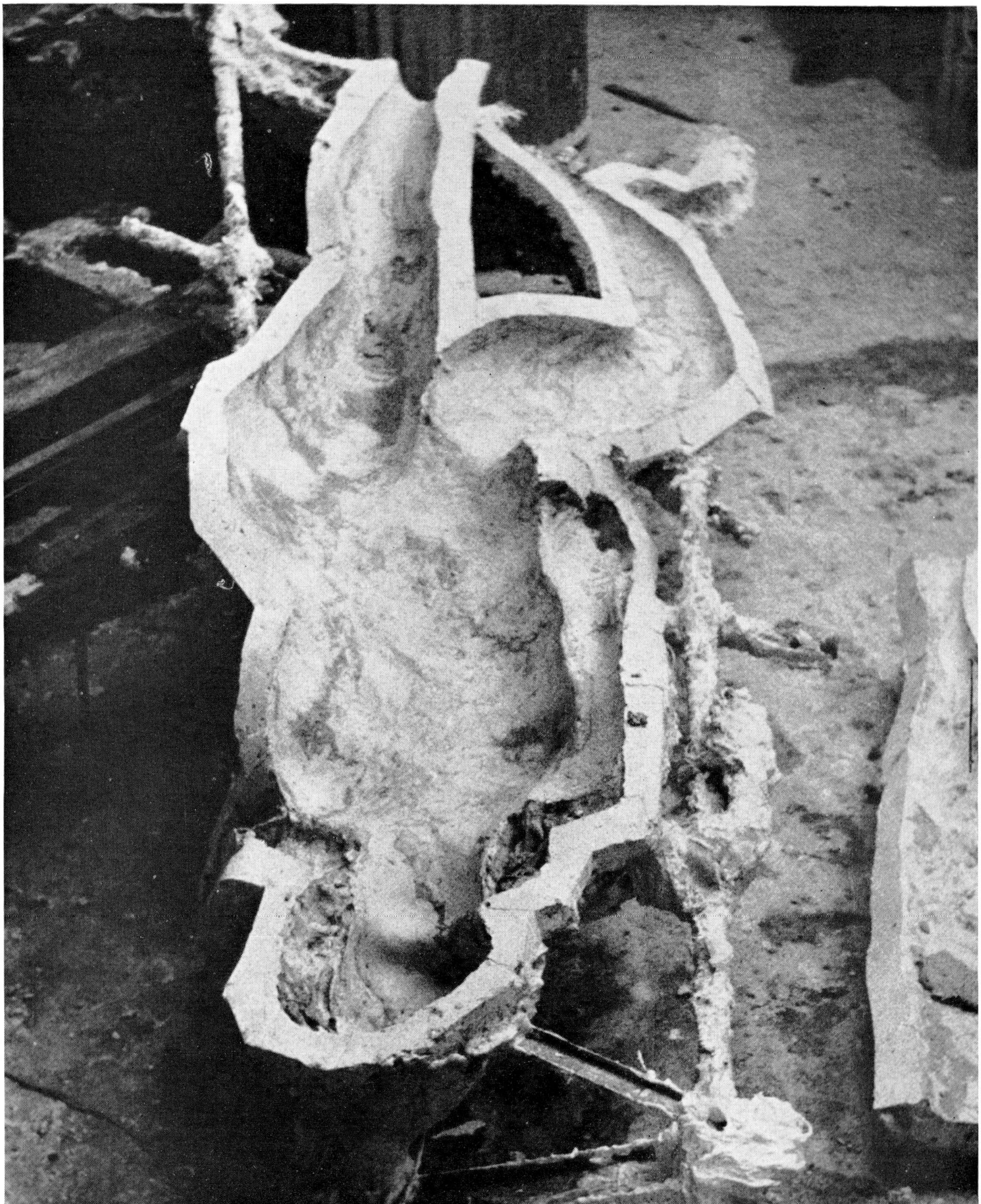

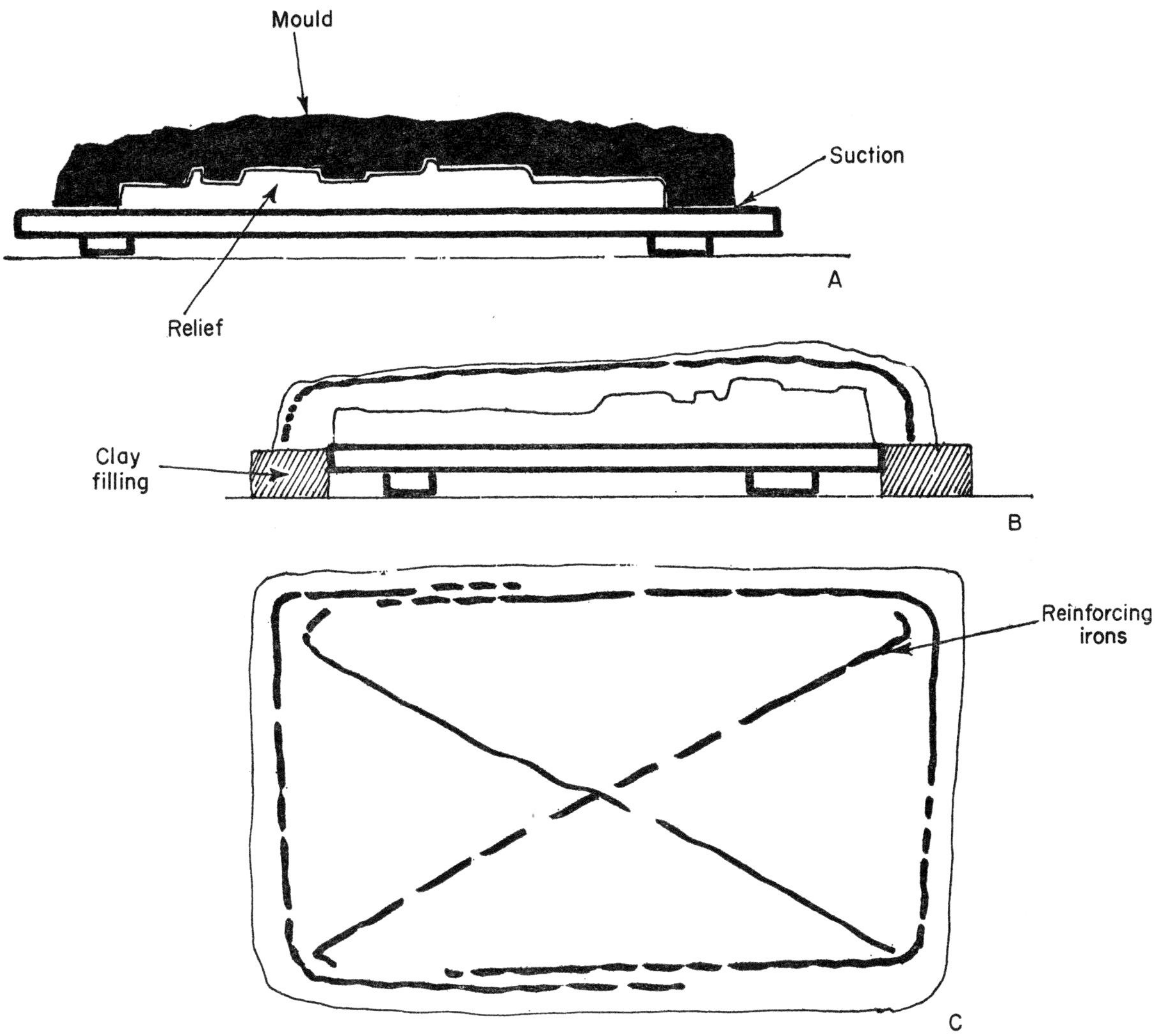

FIG. 1. Diagram explanation for making a simple mould of a relief. The plaster is placed over the clay to give an even thickness reinforced to strengthen it (A) and (B). Suction occurs between the plaster and clay and board, and is released by pouring water between. (B) shows the extension to the base board necessary to make a satisfactory mould thickness, if the board is small. (C) shows in plan, suitable reinforcing of the plaster mould.

The plaster is applied over the entire form, clay and brass. This is best done by mixing small quantities of plaster at first, and throwing carefully from a small plastic bowl over the original. Throw firmly but not too hard. Dip only the fingers into the bowl, and with the back of the hand directing the throw, flick the plaster from the fingers on to the clay. Practice will make this task easy, and with practice and time plaster can be more accurately placed to make the mould quickly. Throwing forces the plaster on to the original surface to give faithful reproduction of surface texture and marks. Always work from the bottom of the work up, to avoid air traps under projecting forms or texture. Use a casting grade of plaster of Paris, not a builders' type.

The first layer of plaster should be coloured to provide a helpful warning when chipping off the mould. The simple addition of powdered water-colour pigment to the water of the first plaster mix will give an adequate warning layer. Do not make this too strong or the plaster will be weakened. After each application of plaster, scrape the top edges of shim. The mould is gradually built up to a required thickness, including the mild steel reinforcing which is placed over the first layer, strengthening the mould integrally. Particular attention must be paid to the seam edges (see Fig. 2). These edges need to be as strong as the materials allow,

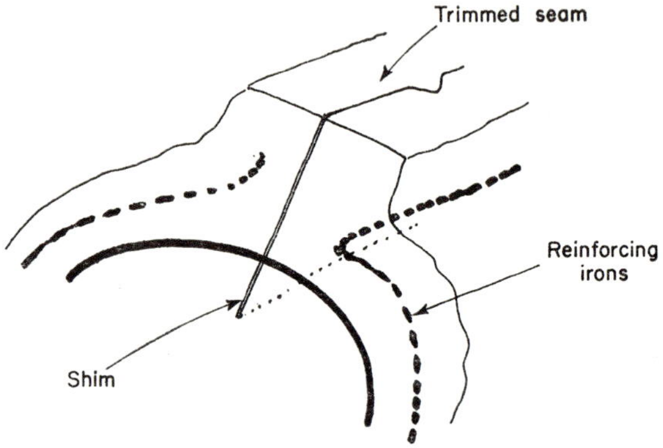

FIG. 2. The section through the completed mould at the seam. The shim is exposed and the plaster trimmed. Reinforcing is as close to the seam as possible.

providing the strength for prising open. A good thickness showing on either side of the scraped brass-shim edge gives a fair idea of strength. When the mould is completed, a suitable thickness built up and strengthened, with shim exposed, a cradle may have to be designed and placed to enable the mould to be easily moved (see Fig. 3). This cradle is fixed with scrim and plaster. The completed mould should stand for at least an hour to harden, before opening, i.e. removing caps by carefully prising. Knives and water are used to open the mould, and to remove the clay and armature. When the caps are removed they must be stood up to prevent warping. Do not lay them down. Prepare the emptied mould (see page 14).

The procedure for waste-moulding using brass shim therefore is:

(1) Design and draw on the model the divisions for caps and main mould.
(2) Cut and place the brass shim.
(3) Mix the plaster and first place coloured layer of plaster mould, not less than $\frac{1}{4}$in. (0·6350cm.).
(4) Scrape the seam edges.
(5) Make and place any mild steel reinforcing.
(6) Build up the plaster mould thickness, which will be uncoloured.
(7) Scrape down and expose seam (brass) edge.*
(8) Allow the mould to harden – 1 hour or more.
(9) Open by pouring water to break suction, and prise carefully.
(10) Remove clay and armature.
(11) Clean out and prepare the mould.

(*If a cradle is required it should be placed after 7).

FIG. 3. The mould showing divisions that make the caps and main mould, reinforcing placed to strengthen the mould, and a cradle fitted to the main mould for extra rigidity and ease of handling.

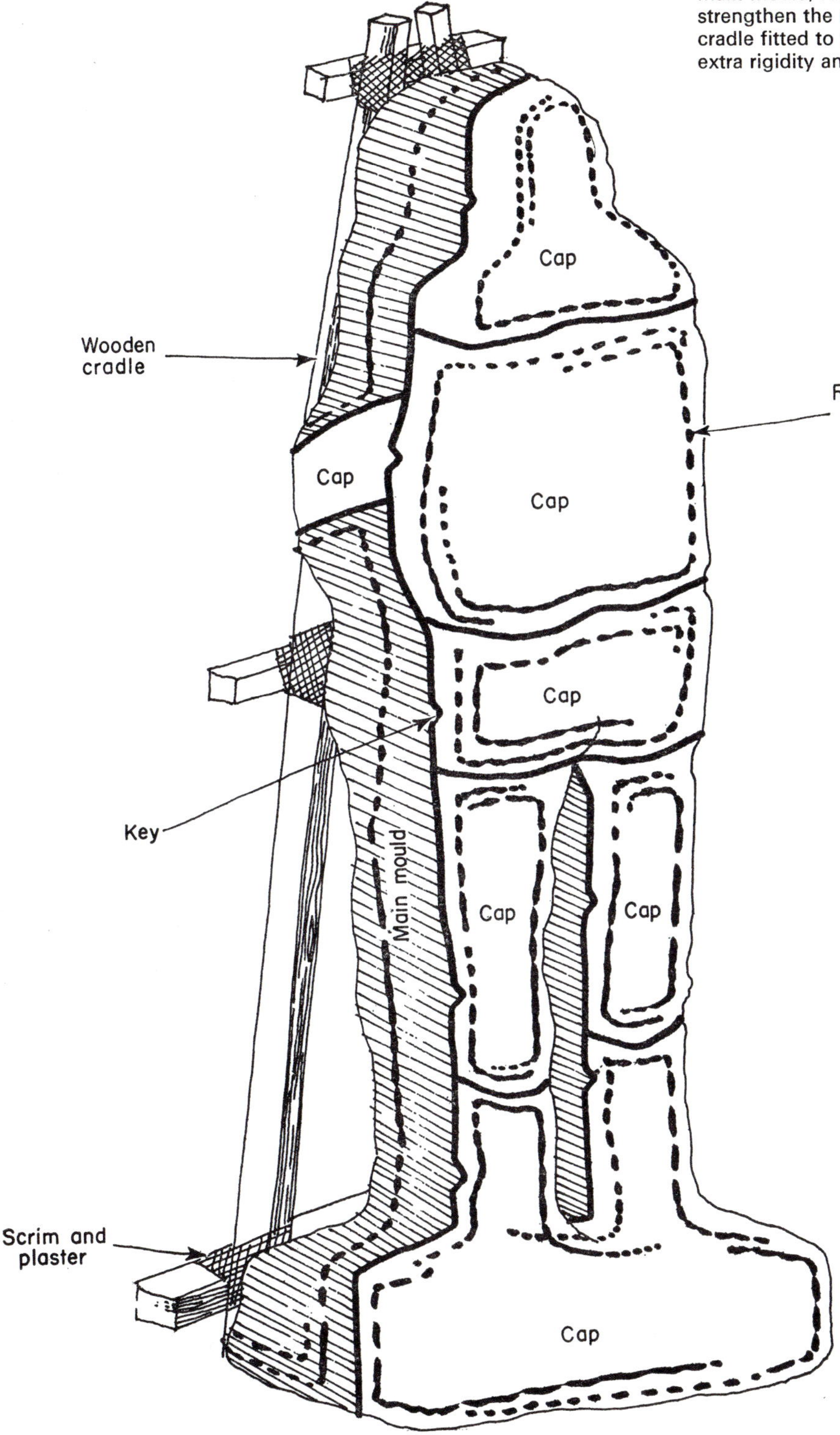

Clay wall

These divisions can be made between caps and main mould, and this technique is used when brass shim is not available, or when the original will not permit shim to be pushed into the surface of the original.

Clay is flattened into strips, and cut into widths equal to the planned thickness of the mould. These strips are placed along the lines drawn on the clay, to make a continuous wall around the cap. The wall is supported by small clay buttresses along the side opposite to the drawn lines. The wallface is made smooth and trimmed. The plaster is placed against this smooth clay wall to make the cap. Only one cap, or one side of a mould, can be made at a time using this method, comprising a coloured warning layer, mild steel reinforcing if necessary, and the plaster thickness. The plaster is scraped down at the clay wall.

When this first side or cap has hardened the clay buttresses and wall are removed. The exposed plaster seam must be trimmed and keyed (with small dimple indentations to locate the caps to the main mould). A parting agent must next be applied to this surface, and for this a clay wash is sufficient. If the mould is a simple two-piece, the next section can now be made. If it is more complex make the next cap in the same way as the first, making another clay wall and so on. The parting agent determines the seam thickness, and subsequent caps and main mould must be accurately placed so that each casts against the prepared seam faces of its neighbour. When the entire clay original is covered with plaster mould, built up with warning layer, reinforcing, mould thickness, and seams scraped and strengthened, the cradle can be built to facilitate handling. Leave it to harden for an hour or more. (The seams will show as hair-lines only, when they are scraped down). When the mould has hardened it can be opened. The caps can be removed together with clay and armature. The mould is then prepared for filling (see Fig. 4).

The procedure for *Clay wall* waste-moulding therefore is as follows:

(1) Design and draw on the clay the main mould and caps.
(2) Prepare the clay wall strips.
(3) Place them with buttress supports to encircle the first cap. Smooth and trim.
(4) Make the first cap, warning layer, reinforcing, plaster thickness, and trim off seam at clay wall.
(5) When plaster has hardened, remove the clay wall and buttresses.
(6) Trim the exposed plaster seam, make the keys and apply a release agent.
(7) Make the subsequent sections of the mould, scraping to reveal seam lines, and allow to harden.*
(8) Prise open carefully, using water and knives. Remove the caps.
(9) Clean out the mould, remove clay and armature (if any).
(10) Prepare the mould for filling.

(*The wooden cradle can be placed after 7)

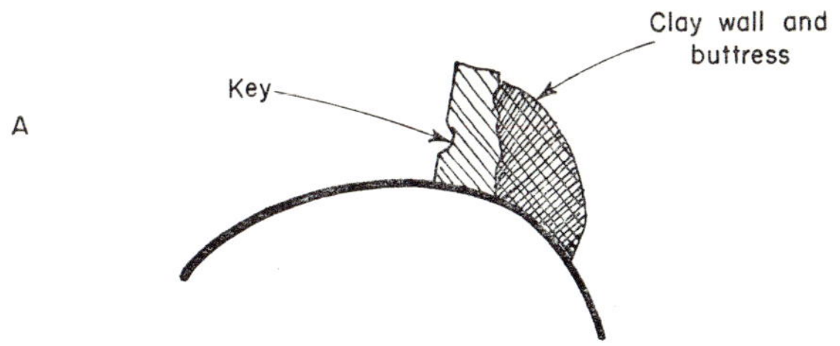

FIG. 4. Description of a clay wall mould. (A) shows the clay wall and buttress, (B) the first mould section made against the wall, the removal of the clay and the release agent applied prior to making the next mould section, as in (C).

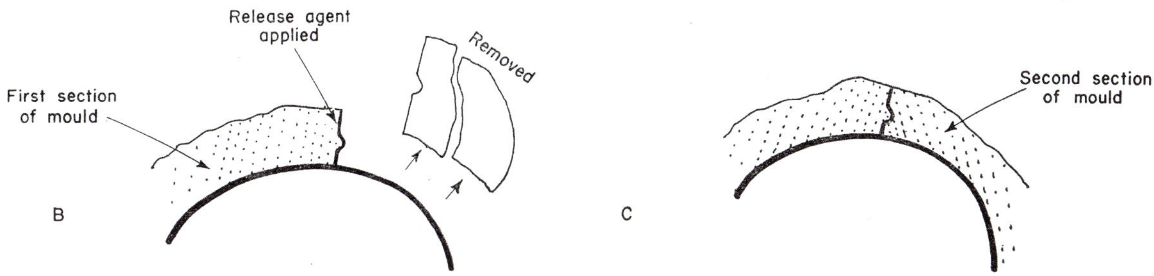

Points to watch in both methods of waste-moulding include making the seams strong, and pouring water into each cap to break any suction between plaster and clay. The water seeping down through the seam breaks this suction. Prising to remove caps without water will almost always result in broken or damaged moulds. When removed the caps must be stood on edge. Now make good any defects such as air bubbles on the surface, broken or chipped seam edges.

When designing the waste-mould, it must be remembered that the entire mould surface must be accessible. On very large forms the division and size of caps should be designed allowing for their weight when filled. When they are replaced, access to the main mould must be retained for even the very ultimate or penultimate cap. It must be possible to reach into the cast to fill the seam, and this means that the size of a cap should be determined by the length of the caster's forearm. The more slender the form the greater the number of caps required. The larger volume requires fewer caps because of easier access.

Chipping out

The technique of removing a plaster waste-mould from the casting, is explained here because it clarifies the title and the nature of this kind of mould. In procedure, of course, it does not occur until after the filling to make the casting has been completed, and the casting cured.

The plaster is carefully chipped away from the casting, and for this blunt joiner's chisels of various sizes and a wooden mallet are used. Commencing at the top of the form in its proper position, the mould is removed firstly to reveal the high points, then carefully chipped away to reveal the whole sculpture. The chisel should be held at right angles to the surface most of the time. It must not be directed across slender forms, in such a way as to break them. The concussion from the mallet must be taken by the mass of the form (see Plate 16). When the mould has been chipped away, any small deposits of plaster can be carefully picked out to clean up the cast. Old dentist's probes are useful for this job. Do not be tempted to prise off large sections of mould, because this may result in pulling off detail, texture or smaller forms. Care must be taken at all times during this operation.

Mould preparation and release agents

The correct preparation of the mould to receive the concrete filling is of major importance. To ensure a good casting, conditions need to be as near as possible ideal for the material.

To achieve the correct chemical action, and effect the proper setting, hardening and cure, both Portland and high alumina cements require the moisture to remain in the concrete mix.

Because plaster of Paris is a porous material, it must be treated in such a way as to retain the correct moisture content in the mix. It is best to carry straight on with filling the plaster mould, as soon as it is cleaned and prepared. In this way it will still be wet. Moulds made and stored for some future occasion will require soaking before being filled.

A mould is properly soaked when water remains visible on the surface, and does not sink into the thickness. Immersion in a large tank of water soaks the mould most efficiently. Failing that, hosing with a spray nozzle is a fair substitute.

The most efficient release agent for preventing adhesion between plaster of Paris and concrete is the fine film of clay deposited on the mould surface from the clay original. If the original is carefully removed and any large or thick deposits cleared out with soft clay dabber, the mould is sufficiently prepared. Apart from soaking this is all that the mould requires. The fine film on the mould surface is such that the smallest mark will chip away quite cleanly. Bloom on the plaster surface from the clay can be scrubbed off when fresh with water and a stiff brush. If allowed to dry, wire wool and a soft wire brush will remove it. It can of course, be retained.

Soft soap and oil applied with a soft brush to a mould surface that has been washed will make a good release agent. The soap should be boiled in water then brushed to a lather all over the mould surface, and left for 20 minutes. It should then be removed with a brush and water. The soaping should be repeated, and finally a fine film of oil applied over the surface (castor oil or clean engine oil will do). A spot of oil placed on the palm of the hand and rubbed well into the brush, thence on to the mould, is sufficient. Soak the mould immediately before filling.

Shellac and wax. If bloom from the plaster must be eliminated from the cast, a preparation of shellac and wax is used. The preparation is tedious. The mould must be semi-dry. Two or three applications of dilute shellac and methylated spirit are made, each coat being left to dry before the next is applied. The shellac seals the mould. It must be dilute to avoid blurring the detail. When the last cast has hardened, liquid wax is applied as an effective release agent. The mould is finally soaked before filling. When the cast has been chipped out, the surface must be rubbed down with turpentine to remove the wax, to prepare the casting for painting or some other patination.

Liquid wax applied in several dilute applications to a semi-dry mould affords a fairly good release. Before filling, the mould must be soaked after the wax has hardened.

A *clay wash* can be applied to a surface of a clean plaster mould. The mould when soaked, will be suitably prepared for filling. It will chip away from the cast fairly well.

Plastic sheeting can be used to prevent adhesion between concrete and simple mould forms. *Waxed paper* is also effective in this way.

The building industry manufacture special mould oils, for treating particular moulds to gain a good clean release, usually from wooden shuttering (moulds). These can be used as release agents from other kinds of moulds that may have a tendency to stick firmly to a concrete casting. Paint, whitewash, lime wash (not for high alumina cement) are all used in mould preparations. The wise man will always experiment with a medium that is unfamiliar, to make sure it will have no harmful effects on the concrete.

Other moulds

There are a variety of other types of moulds from which concrete castings can be made. In most cases they are complex systems, devised to produce more than one casting. It is not possible to give more than a list of these moulds here, together with brief descriptions. More information can be found in *The Technique of Casting for Sculpture.*

The simpler techniques begin with the use of natural materials:

Clay can be used as a mould to produce concrete panels or reliefs. A technique used in some teaching studios, uses clay as a moulding medium. Its malleability and ease of handling are admirable qualities. Items can be pressed into a substantial layer of clay and removed to leave a clear impression. The depth of impression can be varied, and a design can be composed of various pressed elements. What is achieved in this way is, in fact, a mould cavity or negative. This can be elaborated or enhanced by modelling, incising etc. The clay is allowed to become leather-hard, that is hard but not dry. Mould oil or some similar release can be applied if desired. Concrete can be poured or packed into this mould to make a casting. When the concrete has cured, the mould can be soaked and washed off.

Wax is a very useful material in all sculpture techniques. Microcrystalline wax is the most versatile and useful. As a moulding medium its use is very simple indeed. It works most satisfactorily from a clay or synthetic-clay original. All that is necessary is for the original to be dipped several times in and out of molten wax, to receive a substantial covering deposit of wax. A thickness not less than $\frac{3}{16}$in. (0·6350cm.) is necessary; and it will need to be thicker over large forms. When this wax deposit has hardened it can be cut through with a sharp knife and immersed in water. The water softens the clay sufficiently for it to be removed, and the mould surface cleaned. When clean the mould can be filled. The wax mould can be cut easily and the caps designed at the cutting stage. To replace and fix the caps to the main mould is simply done with molten wax, or by welding with a hot steel spatula. Mould oil can be applied lightly to the surface, but it will peel easily away from the casting, if it is warmed.

Plaster of Paris can be used to make a crude but useful mould. This is done simply by covering the original clay with plaster to build a substantial thickness. When this has hardened the mould can be divided by sawing. The clay is then removed and the mould prepared and filled.

Wooden moulds are perhaps amongst the most common in the building industry – shutterings and frameworks are usually of wood, often built *in situ*. These are in effect wooden moulds, presenting large cavities to be filled with concrete. Ways of adapting these in the studio are obvious, and often expedient.

Expanded polystyrene and *rigid polyurethane foam* are both expanded polymers or plastics, widely used today in many spheres such as moulding and pattern-marking materials. They are both resins, extruded with a blowing or foaming agent, giving an expanded cellular material, containing up to 80 per cent air. The resulting virtues are light weight, easily disposed of bulk, and above all they can easily be cut and formed. Mould cavities can be built with them, in the same way as wooden shutterings, although it may be cheaper to fix layers of either expanded polystyrene or polyurethane to the wooden moulds. They can be sawn, glued, pinned etc., and modelled to some degree with hot metal tools. Hot-resistance wire-cutting equipment is the most satisfactory means of cutting these materials, together with sharp knives and scrapers. Special adhesives are required, supplied by the manufacturer or supplier. Other glues will simply dissolve the material.

Both expanded polystyrene and polyurethane will retain a pressed impression, and so this technique plus cutting and modelling, can be used to give infinite variation to mould surfaces and form. Combined with wooden formwork, architectural relief decorations can be made to cover great areas with a wide range of form, texture and effect. This form of decoration – achieved by enlivening the inner *faces* of shuttering and designing patterned and textured surfaces cast *in situ* as part of the building – has been carried to a fine degree of sophistication by William Mitchell in England. Using these techniques he has developed a personal style which is easily recognizable.

Both these expanded polymers can be got rid of from the cast surface by burning with a blow lamp. They volatize and disappear under a naked flame. Paint thinners dissolve them too, so no mould oil or release agent is required except on wood that may be exposed to the concrete filling. Experiments with these materials will undoubtedly prove worthwhile.

Warning must, however, be given about both substances. Masks or a covering for the nostrils and mouth, should be worn when working with them. Expanded polystyrene and polyurethane are granular and dusty and particles tend to float in the air. If breathed in, these particles will remain in the stomach immune to digestive acids, and unlikely to pass through the body system. Polyurethane gives off isocyanitic gas when heated, so care must be taken when using it. The studio should be well ventilated.

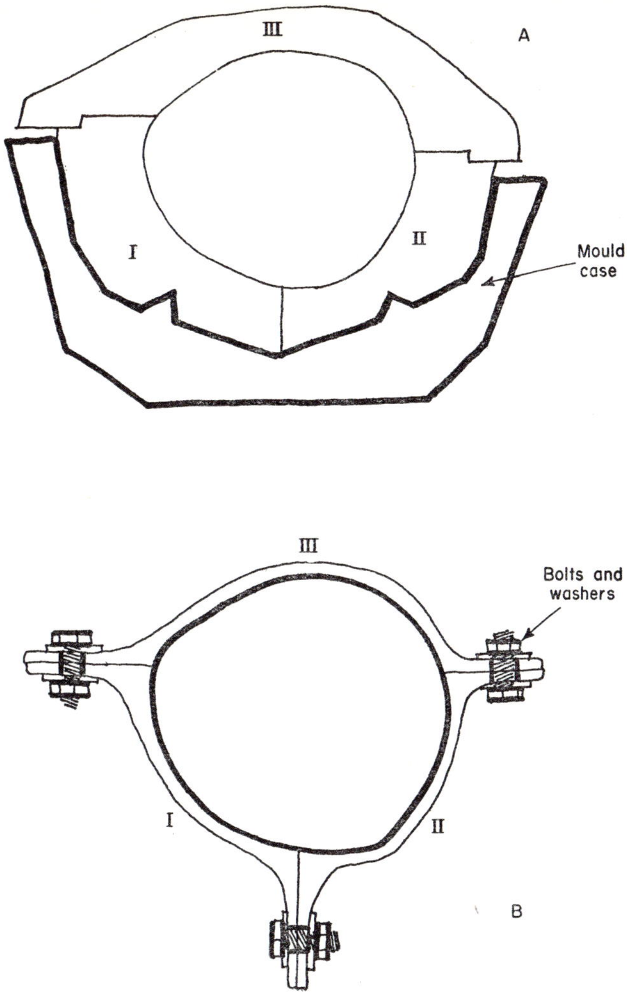

FIG. 5. The principle of piece-moulding is illustrated here. (A) Shows the sections made for a plaster piece-mould contained in the mould case. (B) Illustrates a polyester resin piece-mould and the method of registering and fixing mould sections. Note that three sections are used around the simple form. Two sections could be made with difficulty, three pieces ensure a strong accurate mould.

Polyester resin. This is a thermo-setting resin. It changes from a liquid to a hard insoluble material when a catalyst is added to the basic resin. It is brittle and therefore requires reinforcing, usually given by laminating the resin and glass fibre. It is an excellent mould-making and casting media, with fairly good resistance to weather. It can be made very strong. Simple moulds can be made with it, by building a thickness of reinforced polyester resin over a prepared former (see *Glass Fibre for Armatures*, Lewis & Warring, Tiranti). It has strong adhesive qualities, so proper release must be made between it and the former. Once, however, a mould is made and suitably reinforced, a greater number of concrete castings can be made, with, of course, a suitable mould oil applied to give a good release. Such moulds are often used in building, when a number of identical units are required. Luigi Nervi, the Italian architect and engineer, uses such moulds to achieve some of the beautiful and startlingly fine forms so characteristic of his architecture.

Piece moulding. This is an extremely complicated kind of mould to make, but one that enables a number of castings to be made of complex forms in the round. The principle is to design and make separate pieces of mould for each undercut form. It is taken usually from a hard original, using the *clay wall* technique. It can be made of plaster of Paris, preferably a low-expansion hard plaster. It can also be made of reinforced polyester resin. The pieces of mould, which may be many, are held together by a case or jacket to ensure correct positioning and accurate volumetric reproduction. Much practice, patience and familiarity with the principles of mould-making are necessary to make a good piece mould. Piece moulds are carefully prised from the positive case (see Fig. 5).

Flexible moulds, permitting a number of identical castings, are the most common means of producing more than one cast today (Fig. 6). Gelatine is the traditional flexible moulding media, but it is limited in use, having no storage life, being easily damaged, and restricted to use only with Portland cements. It has been superseded in many applications by a polyvinyl chloride, such as vina-mould, a rubber-like thermoplastic. Thc moulding principle is to make a thickness of flexible mould over the original, supporting this with a plaster of Paris case.

This kind of mould is made by firstly covering the original with a layer of clay, which can be of almost any heat resistant material made non-porous if it is not so already. Over this clay, plaster is built up but divided as in any mould. The plaster case and clay are removed, the clay cleaned out and the trimmed case is placed again over the original. One side or piece of mould is done at a time. The void between the case and original is filled with PVC to make the actual mould. The mould will take in the deepest undercut, and permit castings to be made not only of the various concretes, but of almost any cold-casting material. The mould is simply peeled off the casting, no release agent being necessary. Like piece-moulding however, it is not a skill that can be easily learned from a book, but must be studied in the studio and practised. Initial costs are quite high. A special electric aluminium melting pot is needed to melt the PVC.

Vacuum-formed plastic moulds are another innovation that has grown out of the plastics industry. A thermo-plastic sheeting is softened by heating, and then pulled forcibly over a pattern or former, by creating a vacuum below the pattern. When the pattern is removed after cooling an efficient mould remains. Identical castings can be made for a limited run. Mould oil must be used to give a good release.

Silicone rubber and latex. Both these materials can be used to make effective moulds for concrete castings. Supplied as either liquid or paste or both, moulds can be built up over the original. Dividing the mould by cutting through the build-up or by making sections using the clay-wall technique, satisfactory moulds can be made. Useful as these materials are their high cost is prohibitive. They are liable to shrink, and so their usefulness may be limited in sculpture. The fact that other materials can be used to gain similar results almost excludes them.

Constructed moulds, built *in situ* as a shuttering or when building up a work direct in concrete, offer a wide range of surface and finish. Glass as a mould surface gives a fine, almost polished, surface, possible to achieve in no other way. It is often used for cast bases or plinths for presenting works. Polythene sheet and waxed paper are similarly used to produce cast surfaces peculiar to the character of the mould surface. Again experiment is advised, proceeding with thoughtful caution to allay possible frustration, and will prove expedient and beneficial in the long run.

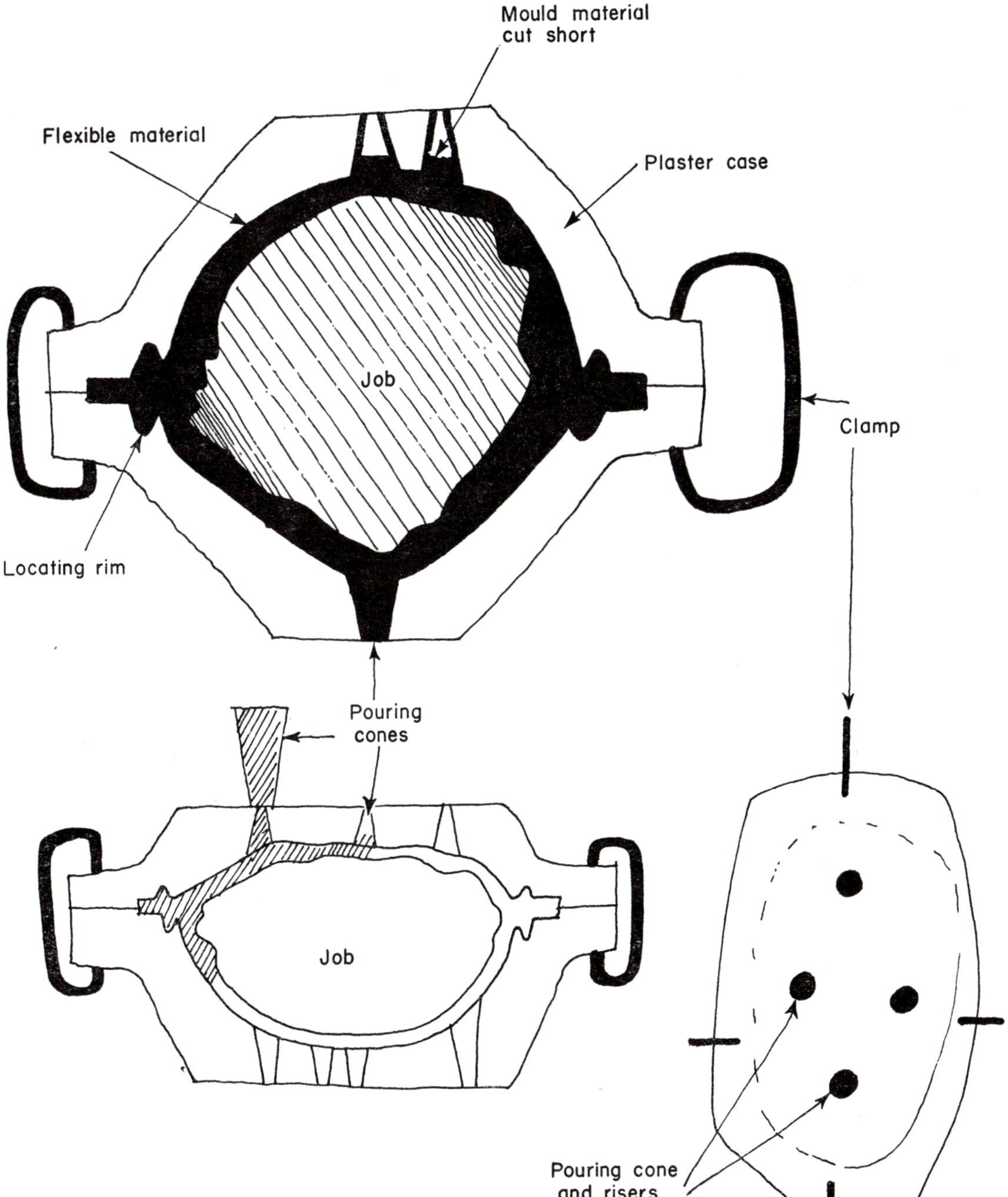

FIG. 6. Diagram of a flexible mould and case. This method can be used for polyvinyl chloride, gelatine, latex or silicone materials.

2 The Concrete

Cement

The nature of cement and its action and importance as a binder, have already been mentioned. Portland cements (calcium aluminium silicate) and high alumina cements (alumina), constitute the two basic kinds. There are many proprietary cements on the market in various countries, mostly derivatives of these two.

Storage

Cements are purchased from the suppliers as fine powders contained in stout bags or sacks, usually of 1cwt. quantities. Storage of the cement is important. Because the material is hygroscopic, attracting moisture from the atmosphere, it needs to be kept dry. It should be stored off the ground in a dry place. Bulk storage does not often concern the sculptor, so storage of large quantities is not usually catered for as on a building site. The problem of preventing deterioration of the material is similar however. If the cement becomes exposed to damp, draughts etc., it is likely to 'air set', becoming lumpy. If lumps in the cement can be easily crushed with the fingers, the cement is unfit for use and should be discarded. Small sacks of cement can be stored by placing them in strong polythene bags, tying the necks tightly to seal off the contents.

Water

Both Portland and high alumina cements become hard and insoluble when mixed to a paste, with water. The proportion of water is important. In principle the effects of water on the mix are as follows:

A. Too much water may weaken the adhesive strength of the cement.

B. Too little water inhibits proper hydration, weakening the cement.

The cement:water ratio therefore must be recognized and remembered when concrete mixes are made. 100lbs. (45·360Kg.) of Portland cement requires at least 25lbs. (11·340Kg.) of water, sometimes as much as 50lbs. (22·680Kg.) to make a workable mix. Water in excess of 25lbs. (11·340Kg.) remains for a while in the set concrete, eventually evaporating to leave voids and channels. These inevitably weaken the compressive strength of the concrete, causing crazing and cracking. As a rule of thumb, the wetter a mix needs to be for workability, the greater the cement content needs to be, and *vice versa*. The water must be clean.

Aggregate

The aggregates are of obvious importance to the mix, because they are its strength. They affect the colour and the texture, and of course the proportions of a mix. They must be clean, free from any contamination, and of an inert nature. The size of the particles, the 'sieve size', can vary from dust up to larger particles of 1in. (2·5399cm.) or more. The ideal aggregate consists of an even graduation from the smallest to the largest particles. The size of the aggregate determines the cement content, by means of its surface area. The larger the aggregate, the smaller the surface area. That is to say the surfaces where particles touch to be stuck together, are less in a large aggregate than a small. The smaller the aggregate

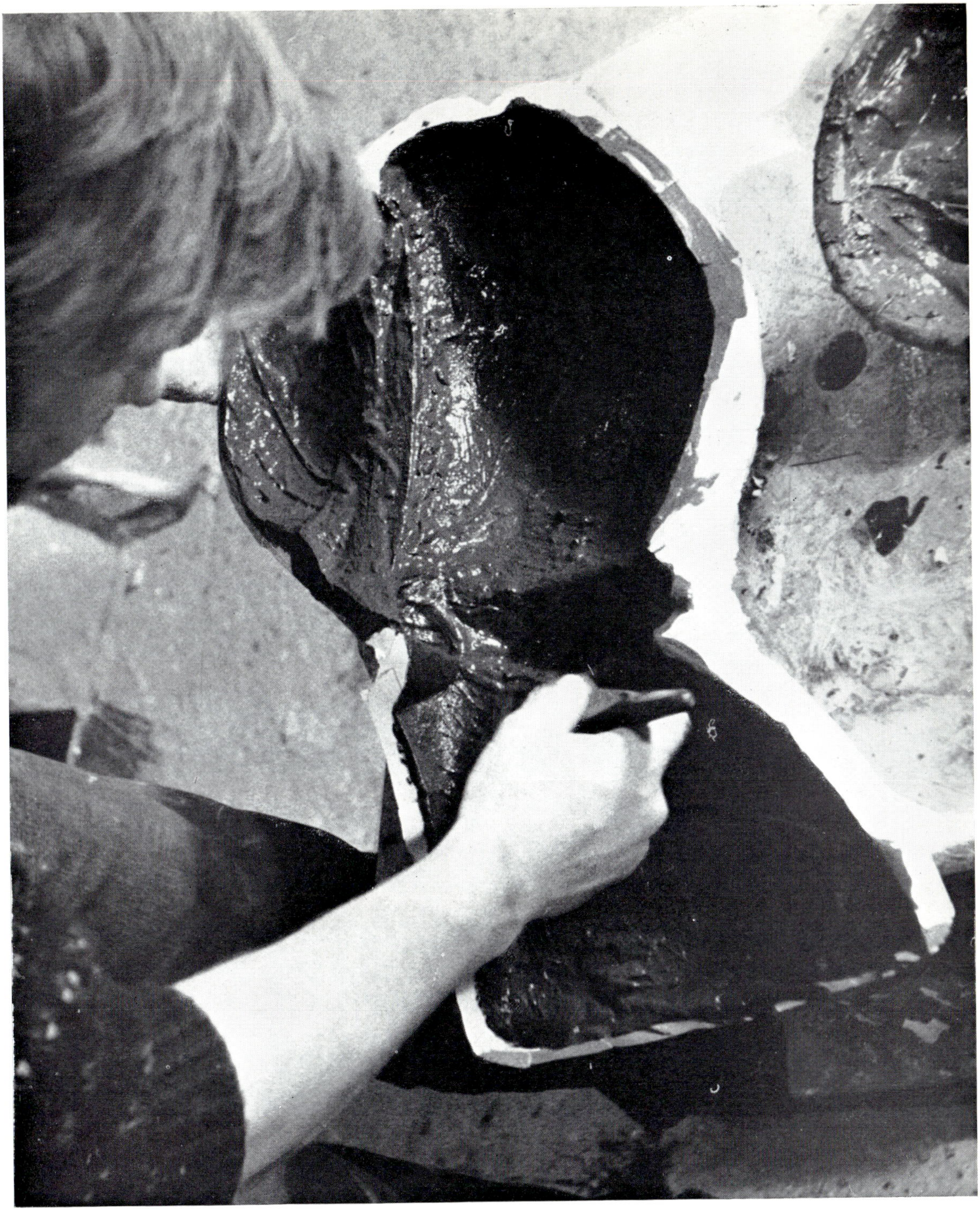

Plate 10. The first layer of wet mix being applied to the prepared mould. This must be brushed well into the detail on the mould surface.

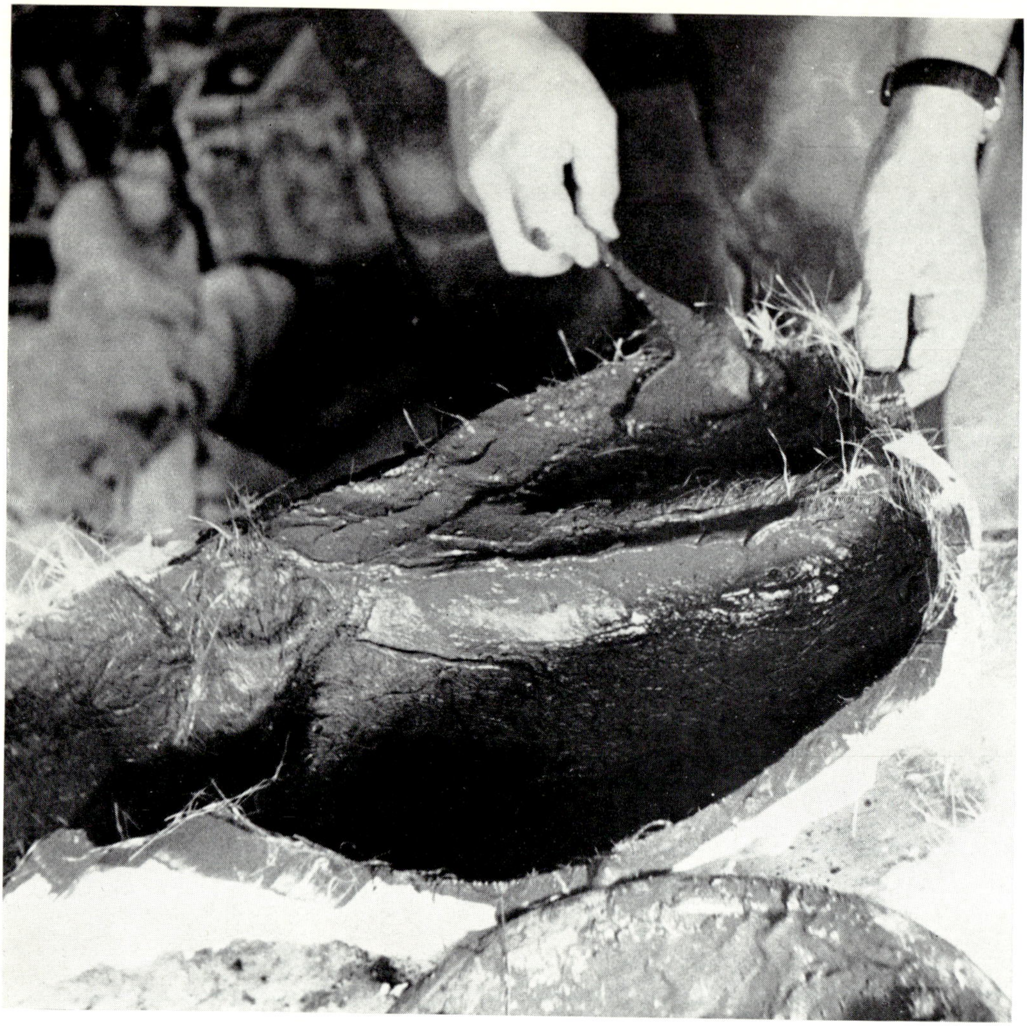

Plates 11 and 12. The build-up of concrete over the glass fibre to form the laminate. This also shows the final treatment to the inner cast surface.

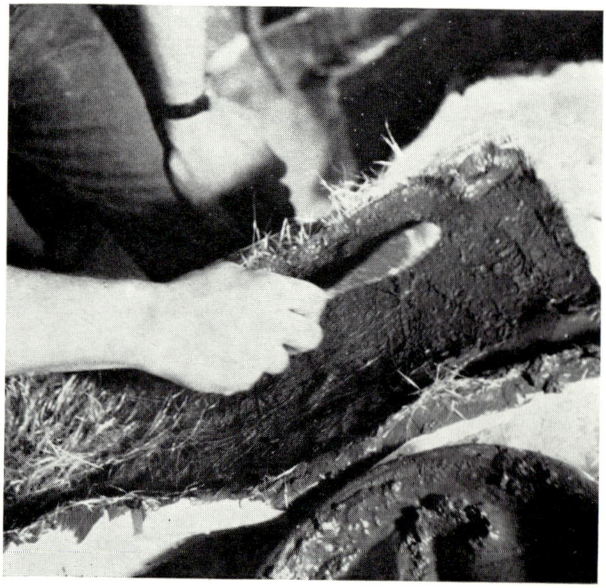

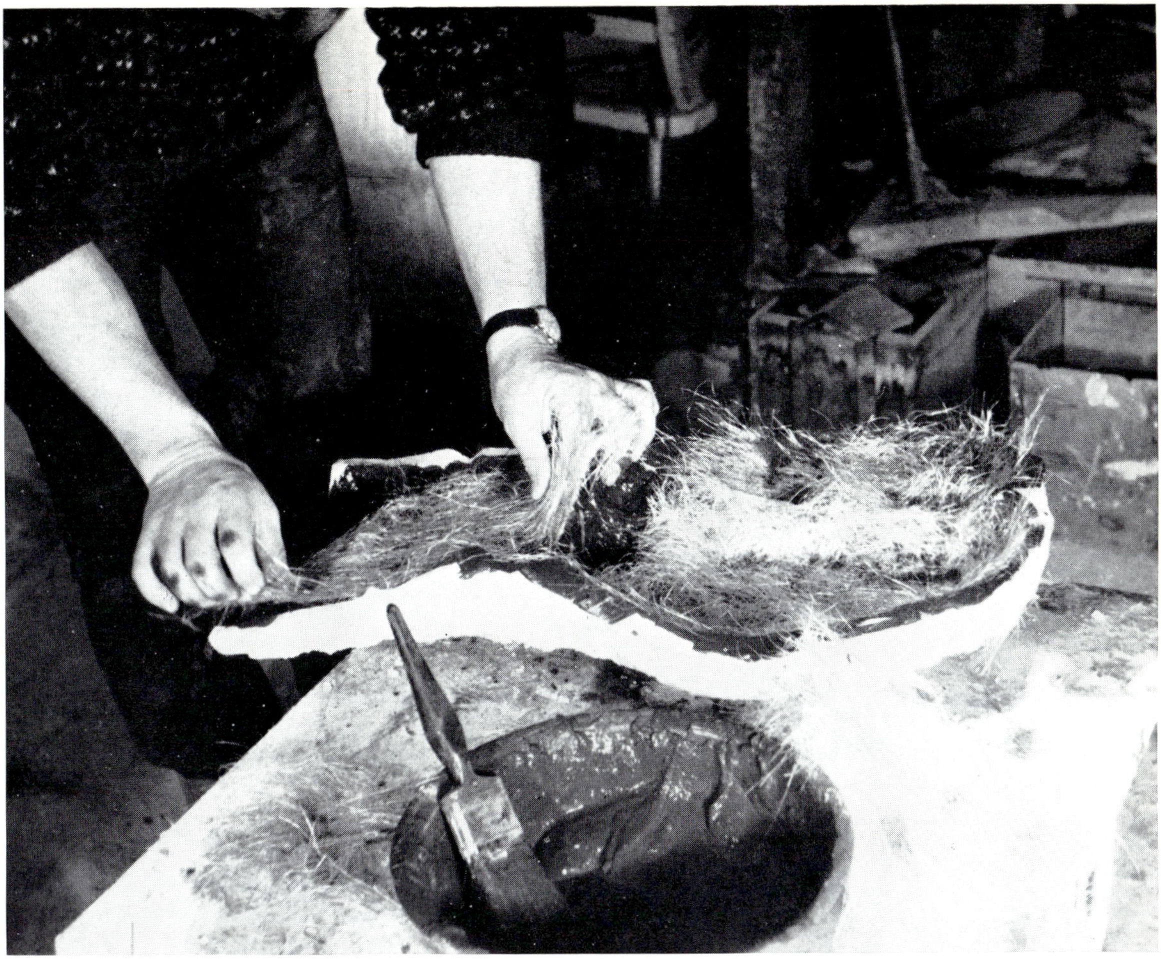

Plate 13. A laminate of glass fibre placed carefully on the first layer of wet mix.

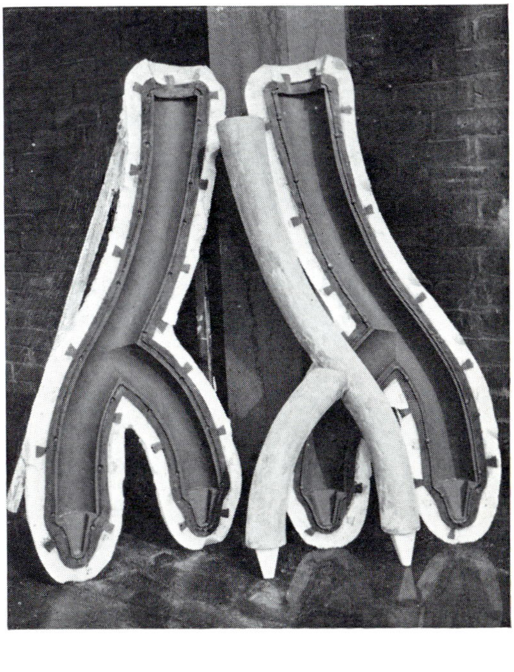

Plate 14. A polyvinyl chloride flexible mould, and the form from which it has been made. The method of fixing the mould to the plaster jacket is clear, as well as the registration keys.

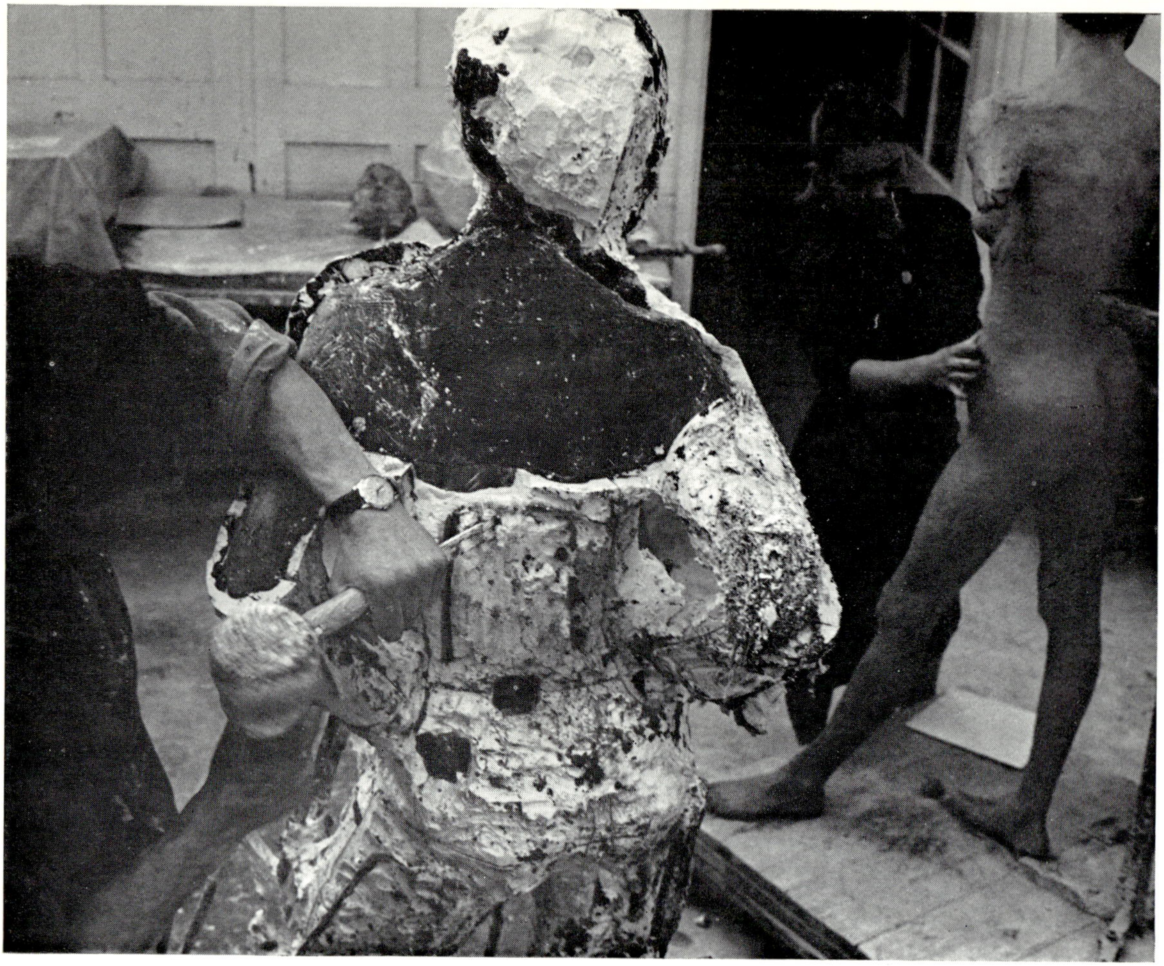

Plate 15. Chipping out a life size figure in a college casting studio.

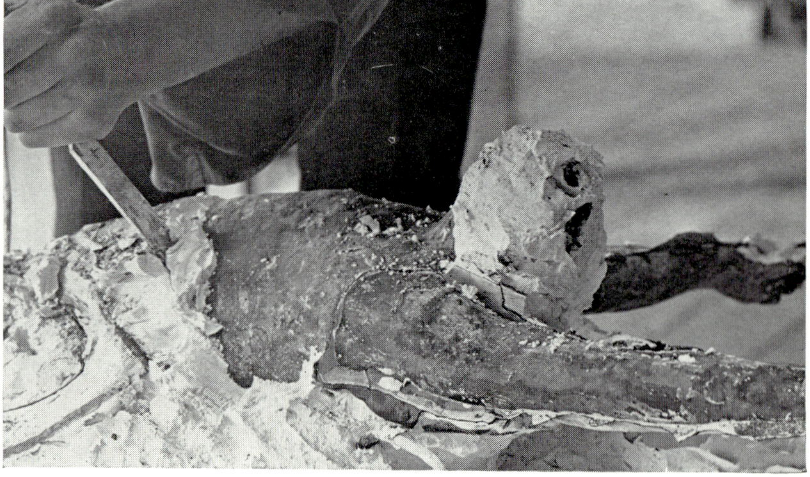

Plate 16. Chipping out, showing the angle of the chisel to the cast surface and the bursting of the plaster mould. The seam flash can be clearly seen. It should be continuous and even.

the greater the surface area, requiring, therefore, a higher cement content. The smaller surface area needs less cement.

Seen in conjunction with the water : cement ratio, the small aggregates, because of their greater surface area, can use a higher water content. The larger aggregates need less water, making a stiffer mix. If for workability more water is needed then more cement should be added, to ensure proper hydration and strength. The smaller or finer aggregates are most suitable for casting because they can be more fluid.

Mixing

The proportions of the mix are determined by the constituents, which in turn are determined by the nature of the final product, colour, weight, strength, etc. The actual mixing is the same no matter what the proportions of ingredients. A 'dry mix' of materials is made first. The ingredients are mixed to an even colour throughout, with no pockets of unmixed aggregate or cement remaining. A poor mixing will weaken the final concrete, and any unmixed aggregate always contrives to be on the surface of a casting, with obvious results. A 'wet mix' is next made by adding water to the 'dry mix'. Evenness of colour must be retained, and the mixture must be consistently wet throughout. Water, if poorly mixed, can act in such a way as to wash out the cement binder, leaving uncovered aggregate and poor concrete. Make both the 'dry' and 'wet' mixes carefully and thoroughly. The appearance, strength and durability of the final sculpture depend on it.

Actual mixing of ingredients for concrete to fill a mould is best done in a mixing box in the studio (Fig. 7). This is a box approximately 2ft. 6ins. (76·197cm.) square, × 10ins. (26·399cm.) deep, lined with plastic or painted with several coats of polyurethane paint or varnish, to make it impervious. A very useful substitute for

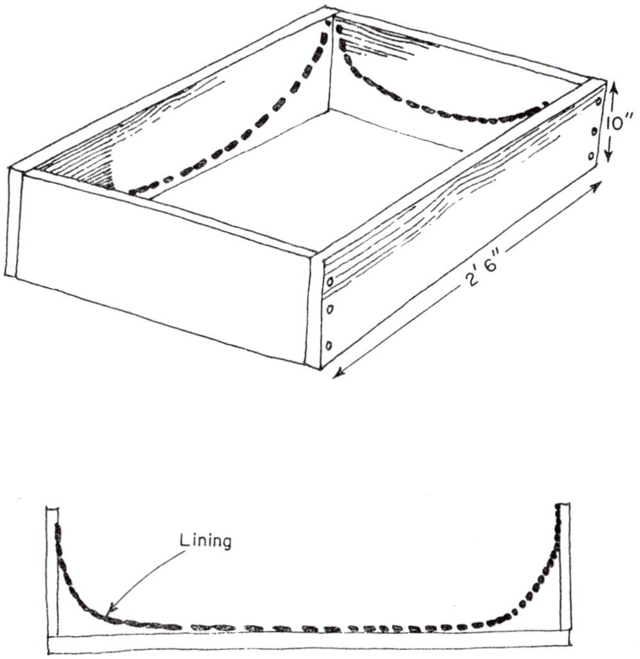

FIG. 7. A wooden mixing box to make both wet and dry mixes. This should be lined with plastic sheet or polyester resin, to seal the wood and eliminate corners to facilitate mixing.

a wooden box is a baby's plastic bath. These are designed to give sufficient depth and access, and are without corners, making admirable mixing vessels. Use one box or bath for dry mixing, and another for wet mixing. Sculptors generally mix with their hands, rather than with a trowel or shovel. Serious harm can occur to the skin however, so it is wise to wear thin plastic or domestic rubber gloves, or to apply a strong barrier cream to protect the skin.

Setting – hardening – curing

When water is added to the concrete mix, the process of hydration begins, causing the cement to solidify, and bind the aggregate. The initial change from a fluid to a solid condition is known as the 'setting', not to be confused with 'hardening' which is the strength development time, after setting. 'Curing' is the period of maintaining the development of strength in the solid, by prolonging favourable conditions. Hardening, and therefore curing, will cease as the solid dries out. Proper cure will depend upon the retention of moisture in the concrete. The development of strength in the solid concrete goes on indefinitely, if conditions are favourable. The following chart can be used as a rough guide:

	Setting Time	*Hardening Time*	*Curing Time*
Portland cement	10hrs.	3 days	21 days
Rapid hardening Portland	10hrs.	24hrs.	7 days
High alumina cement	3hrs.	7hrs.	24hrs.

It will be seen that high alumina cement has the quickest setting and hardening times. It 'cures' sufficiently in 24 hours to be used. Of course if it is kept under favourable moist conditions the cure will go on very slowly developing strength. This fact applies to all types of concrete. In terms of actual compressive strength results shown in pounds per square inch, have developed as follows:

	24hrs.	*3 days*	*28 days*
Portland cement	—	1,500	4,000
Rapid hardening Portland	500	2,300	5,000
High alumina cement	5,600	7,300	8,500

These figures are given as a guide only and are not definitive, because of the strength variability of aggregates.

In practice this means that when a mould has been filled, both the filling and mould must be covered with wet cloths and plastic. High alumina cements must not be allowed to dehydrate, and the best results are achieved by covering completely with saturated cloths, sealing this moisture in by wrapping tightly with plastic sheeting for 24 hours. Portland cements require less moisture than high alumina cements, and so are wrapped in wet wrung, but *not* saturated, cloths. They can be allowed to dry out slowly according to the chosen hardening and curing time. The longer they remain wet the stronger the concrete. It is best not to leave a mould uncovered after filling or during the filling process. Crazing and cracking will occur if the concrete is allowed to dry out too quickly and, therefore, shrink. Little or no volumetric shrinkage will take place if the concrete is properly cured.

Aggregates

Because of its finite nature in terms of the image and completeness of form, sculpture develops different needs for aggregates and mixes, from those normal in the building trade. Aggregates are therefore divided to give two categories, fine and coarse. Fine aggregates are used on small works or subjects which are fine in form and need to be light weight, and which require accurate surface reproduction. By their very nature coarse aggregates permit bulky forms that require weight and compressive strength plus the possibility of carving and abrading to achieve particular surface and effect.

Fine. Fine aggregates, of a sieve size equal almost to the cement powder, include:

Pumice – an imported material
Brick dust – crushed brick
Marble dust – crushed marble
Silver sand
Vermiculite
Grog – ground ceramic
Pulverised fuel ash
Stone dust – crushed stone of various kinds
Natural fine sand – various colours from various parts of the country
(Cement 1 :1 with any of these materials)

These fine aggregates are mixed in a 1 :1 ratio of cement and aggregate. In this way the surface area of the aggregate is covered, and a fairly wet mix can be made. This is sometimes applied to the mould surface with a brush.

Coarse. Coarse aggregates of sieve sizes varying from dust to $\frac{3}{16}$in. (0·6350cm.) or more include:

Pumice – imported material
Grog – crushed ceramic
Natural stone – crushed marble
crushed granite
Gravel
Crushed concrete
Crushed brick
Granules of plastic
Clinker

These aggregates are usually mixed by sculptors in the ratio of 1 cement : 3 aggregate. This permits good coverage of the surface area of aggregate, plus a good quantity of 'fines'. The 'fines' are the small particles of aggregate and cement that fill any voids between the larger pieces of material, to make a good dense concrete. Coarse aggregates can be much larger than $\frac{3}{16}$in. (0·6350cm.), but in sculpture these larger pieces are usually placed carefully into the mould, to be exposed on the final surface. It is wise to experiment with any proposed mix to make certain of the quality of ingredients, and the nature of the final material.

Fibrous. Included amongst the aggregates must be the various fibres. In old Roman mortar, horse hair was used as a fibrous additive, to give some resistance to breaking of the final set material, should there be any great strain. In effect the fibre afforded an additional crosslinking in the solid. In 1959 Sydney Harpley and the author experimented with various fibres to give this kind of strength to a concrete casting. It was found that glass fibre, in the form of chopped strand mat, gave the necessary strength. This mat is made up of random placed fibres of about 2ins. (5·0799cm.) in length. Strength is given in many various directions, unlike woven fibres that offer support in only two directions. These glass fibres were used in high alumina concretes. Portland cements are highly alkaline in their setting stage and tend to attack most fibrous additions. Glass fibre is the most successful of those used with Portland however, having so far withstood long period tests. Terylene can be used in the same way as glass fibre. Whatever is used must be well impregnated with cement.

The technique evolved is that of laminating concrete and glass fibre matting, to build a substantial thickness. On some works it is best to include the fibre separated in with the concrete mix, making a kind of dough, which is fairly easy to use. Concrete mixed 1 cement :1 fine aggregate is the most suitable for this kind of casting, fibre being added to the mix to make a dough, or built in laminates with the concrete.

Colour. The basic cements referred to so far are the most common. There are other varieties, mostly derivatives of these cements which are available in various parts of the world. In all cases there are distinct possibilities regarding colour. Portland cements are a pale grey, apart from the special white Portland, which is the most expensive of the range and normally used to simulate stone finishes for facing slabs etc. High alumina cements are almost black in colour and cannot be mixed with Portland cements. The result will only cause a 'flash' set and a weak solid. The various ordinary Portland cements can be mixed. They can be purchased in a number of colours and colouring pigments can be added to them. There is therefore a fairly wide range of possible colour inherent in the material. The qualities combined with the colour of aggregates widen the range further.

Aggregate colour is best exploited by exposure. In his book *Exposed Concrete Finishes*, J. Gilchrist Wilson describes in considerable detail the various techniques used to achieve these finishes. They include the processes of brushing and washing, tooling (basic carving techniques), grinding and polishing, aggregate transfer, sand and shot blasting. These various headings are self explanatory, and as the skills are explained in full by Mr Wilson little description is needed here. Suffice it to say that the concrete must be well compacted and dense, with the aggregate reasonably near the surface. If this is not done it will not be easily exposed, consequently a sculptured form may suffer by excessive grinding or carving. If it is not compacted and seams are not well filled, these flaws will show up when the aggregate is exposed.

Aggregate transfer is a particularly interesting technique. Seen in conjunction with the use of expanded polystyrene or polyurethane on shuttering, it can provide attractive possibilities. Selected aggregates placed by hand, are glued to the shuttering or mould surface.

Water-soluble glue must be used. The concrete when poured, compacted, set, and hardened picks up the aggregate, transferring it from the panels. When the shuttering or mould is removed water is used first to dissolve the adhesive. Various items, motifs of various materials – non-ferrous metals, resins, glass, ceramic, mosaic, enamel – can all be integrated into a design by this transfer technique.

Most exposed aggregate techniques are devised to remove the 'fines' (the fine particles of concrete that fill the interstecies) from the surface. Brushing and washing is done before the concrete hardens. Some tooling is done at this stage too (i.e. combing and scraping). Grinding and polishing, and blasting with grit or shot is done at a prescribed time during hardening.

On very large works such as those typical of William Mitchell, or the large concrete fountain at Manchester by John McCarthy, the nature of the concrete mix is worked out by civil engineers. They work according to the specific requirements as to colour, surface texture etc. of the designer, and draw up the work method, procedure, and mix to be used by the contractor. The reasons for this are obvious. Engineers are trained in the practical uses of building materials, not only in connection with the manifestation of a design, but also according to safety factors, resistance to specific forces, to soil, atmosphere etc. Particular requirements demand special treatment. Surfaces and forms can be determined in consultation with engineers. Advice regarding surfaces and specification can be sought from The Cement and Concrete Research Association, who have an advisory and research station willing to provide specific information regarding cements and concrete. Tapping the knowledge and unique experience built up at this establishment will always prove worthwhile.

Reinforcing

Integral tensile strength is given to concrete by adding reinforcing. Mild-steel rod is the most common because it is more ductile than other kinds of steel, and can be shaped by bending to fit accurately into the mould. Bearing in mind the usual complexities of form in sculpture, this factor is of great importance. Mild-steel rod, bar, or expanded metal, offer a range of reinforcing sufficient to cover most sculptural requirements.

The mild-steel should be free of oil, therefore preferably not bright. Rusty mild-steel, providing any scaling is removed, is the best. The rusting enables the concrete to bind well to the reinforcing, acting as a key to give grip. If new bar or rod is used it should be cleaned of any oil, and the surface roughened to give a good grip for the concrete; it should not be painted. Concrete must completely cover any mild steel reinforcing. If it is not covered completely the mild-steel will oxidize (rust) further and lose its strength. This is certainly the case if reinforcing is left exposed at all. It will rust along its length and the concrete will lose any strength from the strengthening it should provide. If, for one reason or another, part of the reinforcing is left exposed, it must be painted with a protective paint.

In the past, using the larger aggregates, castings were reinforced by

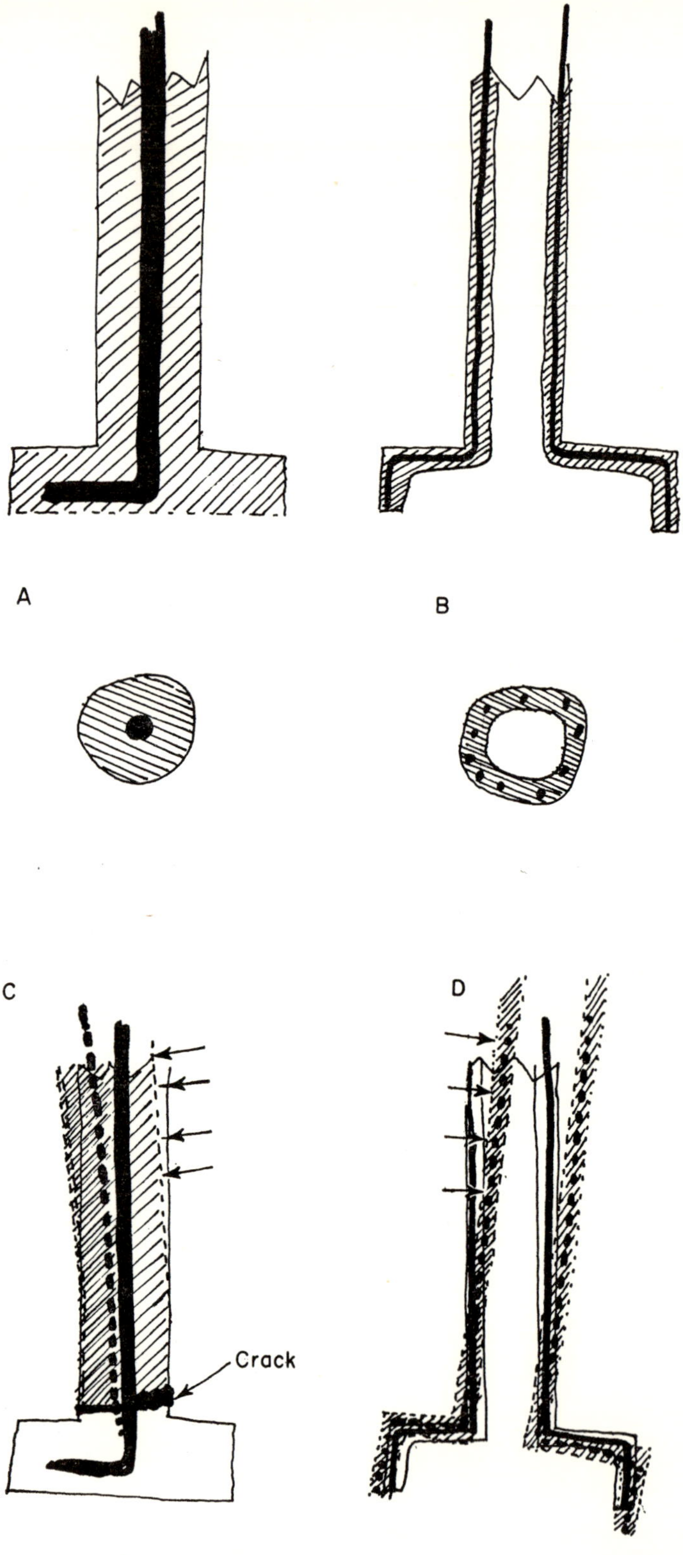

FIG. 8. REINFORCING. Mild-steel reinforcing is best placed as in (B) to form a gaiter support around the cast thickness. The support as shown in (A) provides no strength at the outer surfaces and does not prevent cracking. This method was used to provide doubtful strength to solid castings.

inserting large diameter mild-steel bar at the centre of the form, and the form was filled solid about the bar. This procedure, apart from making an exceedingly heavy job, afforded very little reinforcing to the concrete, and none of the extra strength necessary to ensure durability, and longevity for the sculpture *in situ*. Usually tensile stresses are taken on the outer surfaces of a form, according to the source of the stresses. Reinforcing is therefore required at the surfaces likely to be put under stress (Fig. 8). Working from the knowledge that a tube is stronger than a rod of the same diameter and material, and that a hollow sphere is stronger than a solid of the same dimension and material, reinforcing should be placed in a kind of gaiter principle. Narrow gauge mild-steel rod approximately $\frac{3}{16}$in. (0·6350cm.) placed strategically in a casting permits the strongest result. The narrow gauge permits bending to fit the mould, and also a complete covering by the thickness of concrete used in the castings (which are not made solid).

High-tensile steel rod and wire are used to give great strength to pre-stressed concrete, but are seldom practical in the average sculpture studios. Their low ductibility makes them difficult to bend, and consequently limits their use in sculpture.

3 Castings

The actual techniques of casting in the studio, and of filling the mould to produce a concrete positive, are basically two:

1 The technique for filling with large aggregate mixes
2 The technique for filling with fine aggregate mixes

Before describing these skills in detail, some explanation must be given of the casting principle used in the building industry generally. These are the processes employed by the contractor to produce cast work *in situ*, utilized by sculptors and designers:

A cavity is prepared to receive the filling. This may be a wood shuttering perhaps, a fibre glass reinforced polyester resin mould, or a vacuum-formed plastic mould. The cavity will be dressed, and any additions to the inner faces of the shuttering placed. Mild-steel reinforcing will also be made and placed. The specified concrete mix is then poured into the prepared mould. In the process of pouring air is taken into the mixture of concrete, and the air becomes trapped in pockets. To get rid of this air trapped in the concrete or between the filling and the mould surface, the concrete is vibrated. Vibration makes the air rise and escape, thus compacting the concrete to give a dense solid. Because of this vibration the 'fines' take their proper place around the aggregate, giving also a faithful reproduction of the mould surface. It will be obvious from this brief description that the vibration or compacting of the poured material is very important.

The method of compacting by vibration varies according to the situation and nature of the work. Hand compacting known as 'tamping' is the method most familiar to the sculptor. This is achieved by inserting a rod, dowel, or similar object to release any air, followed by patting with a 'tamper' to ensure dense consolidation. Most people have watched builders tamping layers of concrete for the foundations of a building, or when putting down a drive or roadway. The tamper is usually a wooden plank used on edge. Mechanical vibrators are inserted into the poured mix to compact larger volumes, and these machines can often be seen operating on building sites. Vibrators clamped to the side of shuttering are also used. Kango hammers (pneumatic hammers) are also used. Electric drills with a cambed bit can be used successfully in the studio (Fig. 9). Vibrating tables are used in pre-cast workshops, where much finer tolerances can be worked to than on a building site.

The sculptor must appreciate the necessity for compacting concrete, to ensure the best possible results for his work, for on this depends its durability and strength. Compacting may be achieved in less violent ways for moulds of such delicate nature as those of plaster of Paris, but it must be achieved.

Builders' methods may appear crude, but much can be learned from them. The most expedient technique giving the best result should be sought after, and the cruder methods may suit some forms better than others.

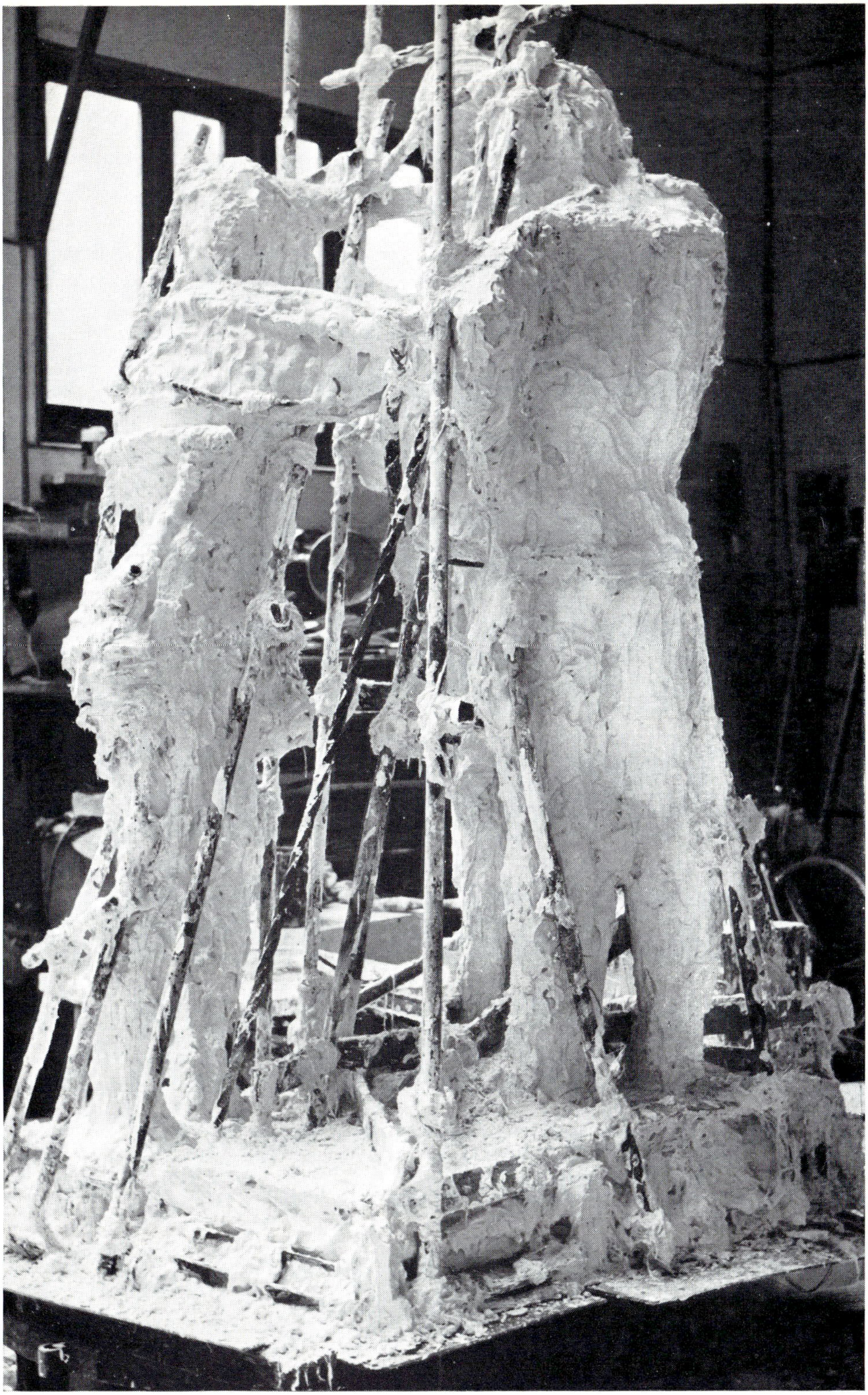

Plate 17. A plaster waste-mould being made on a group of three life size figures. The iron reinforcing and cradle can be clearly seen. Note also the scraping to reveal seam shims.

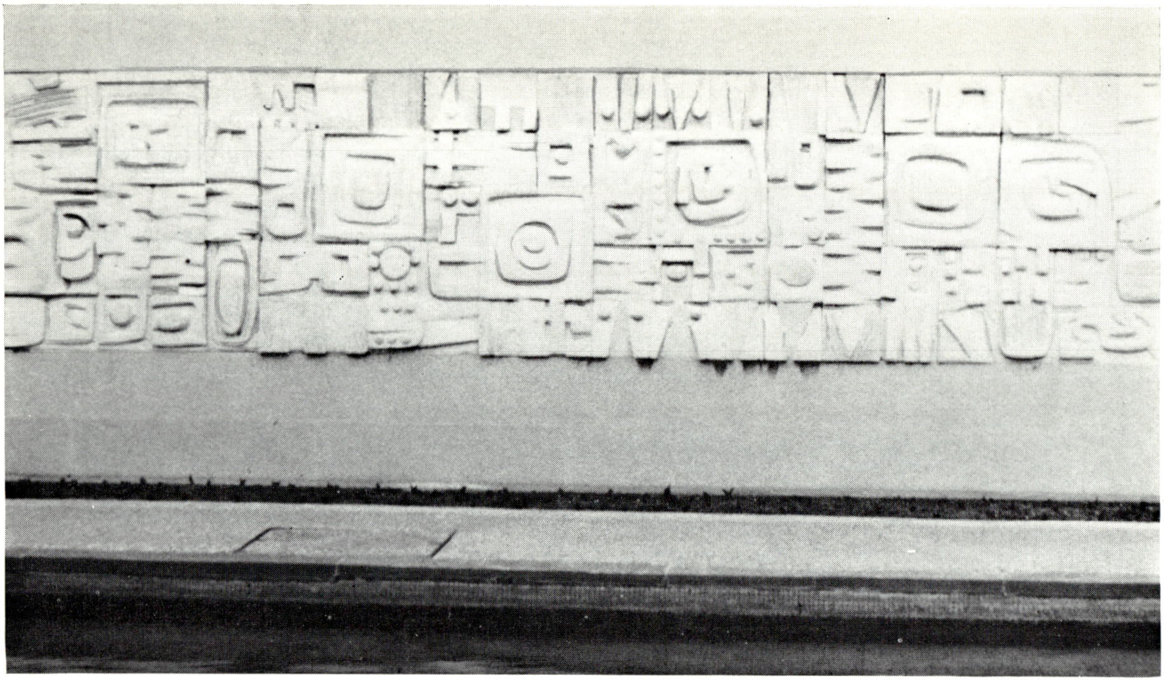

Plates 18 and 19. Decorative wall (which is approximately 20ft. long) at Harlow New Town, by William Mitchell.

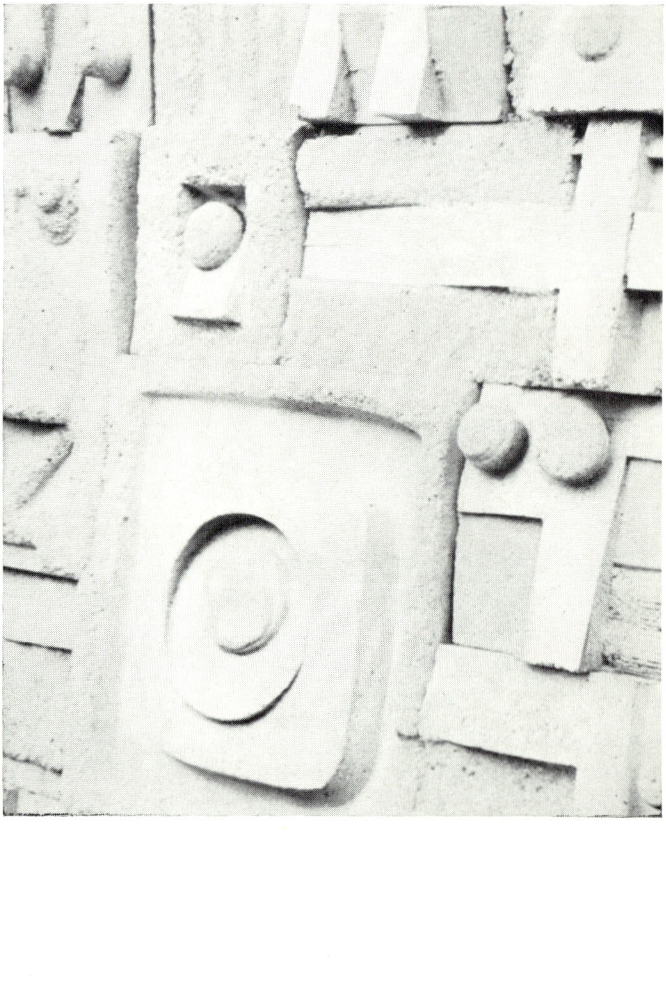

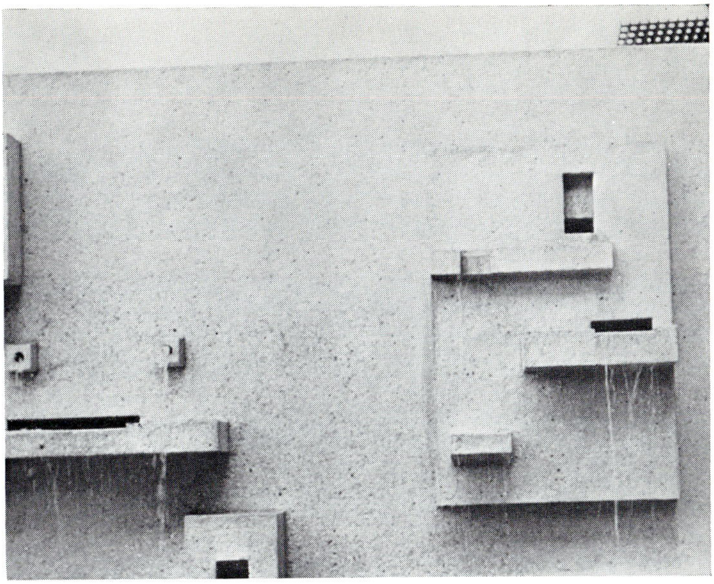

Plates 20 and 21. Fountain in cast concrete by John McCarthy at CWS Building, Manchester. Cast *in situ* by the contractor to the design of the sculptor (11ft. x 60ft.). The plumbing designed and placed in the shuttering.

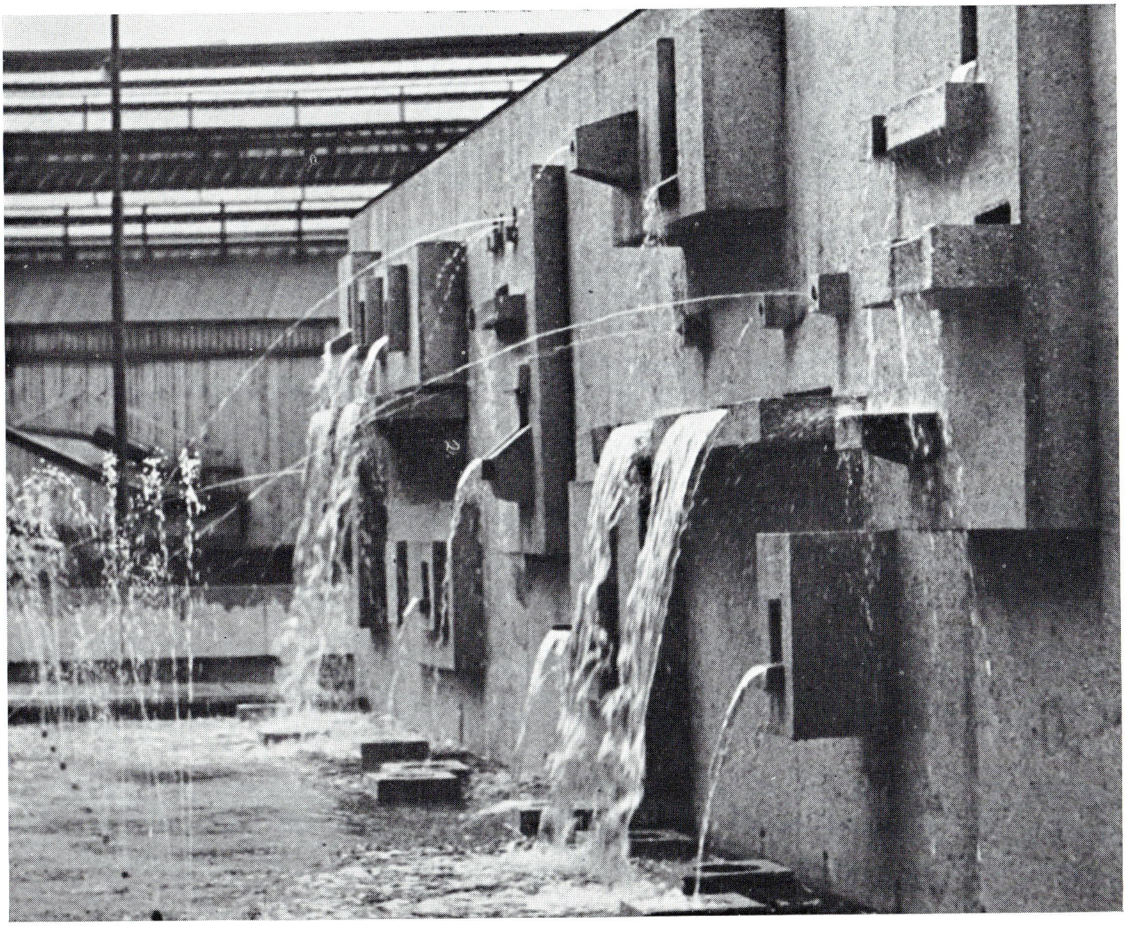

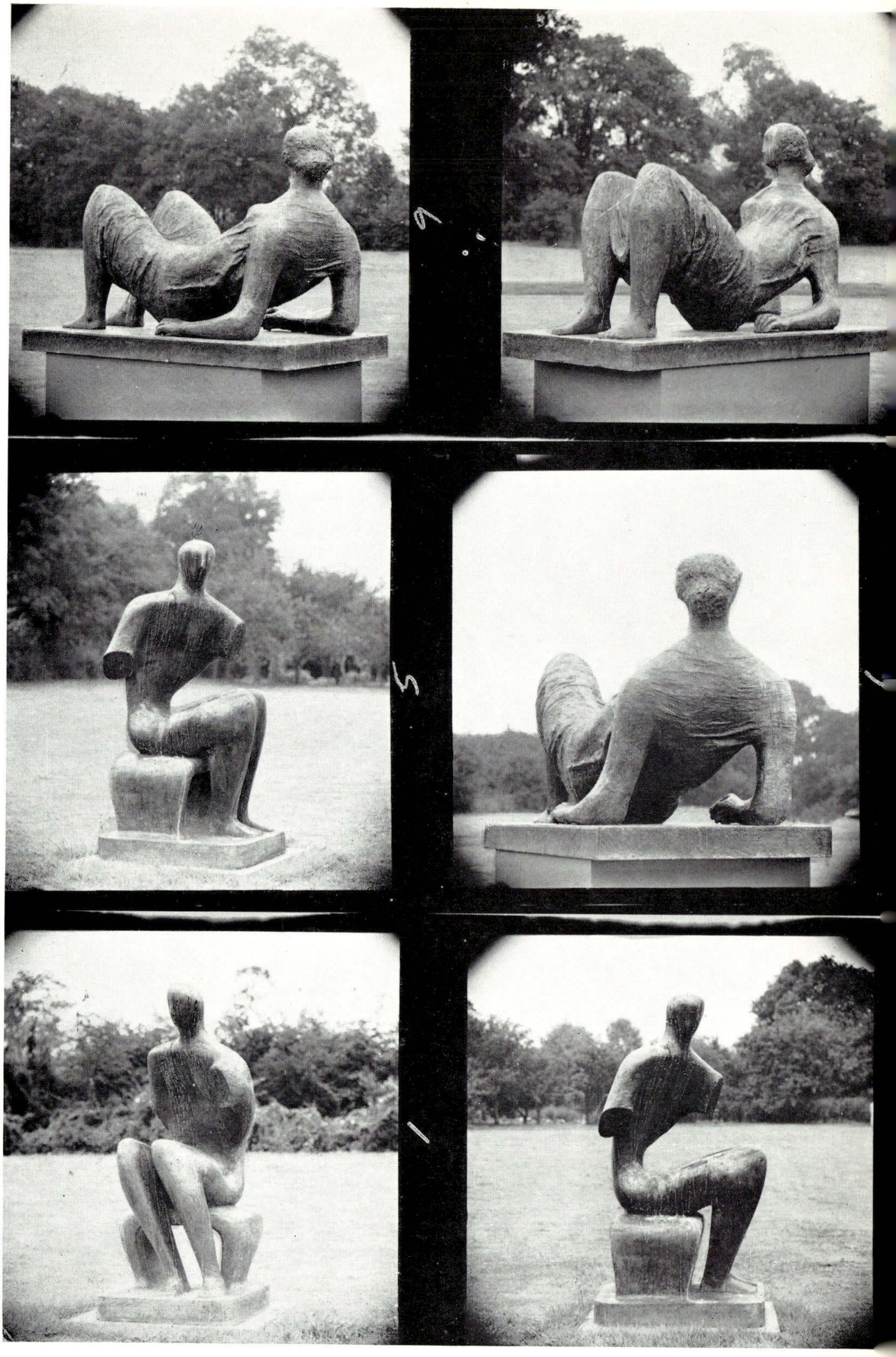

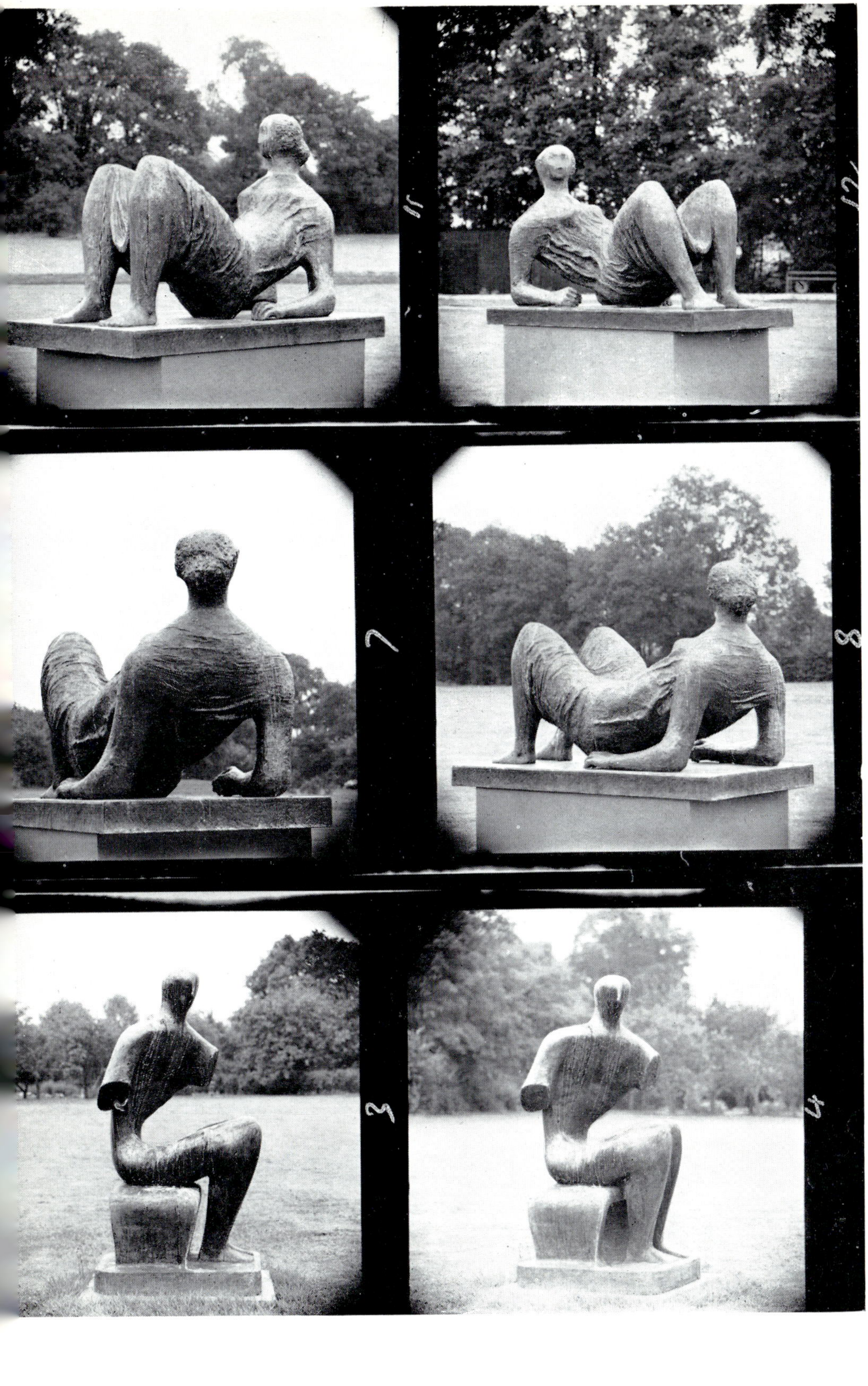

Plate 22. Two sculptures by Henry Moore, cast in aluminous cement and placed in his garden at Much Hadham.

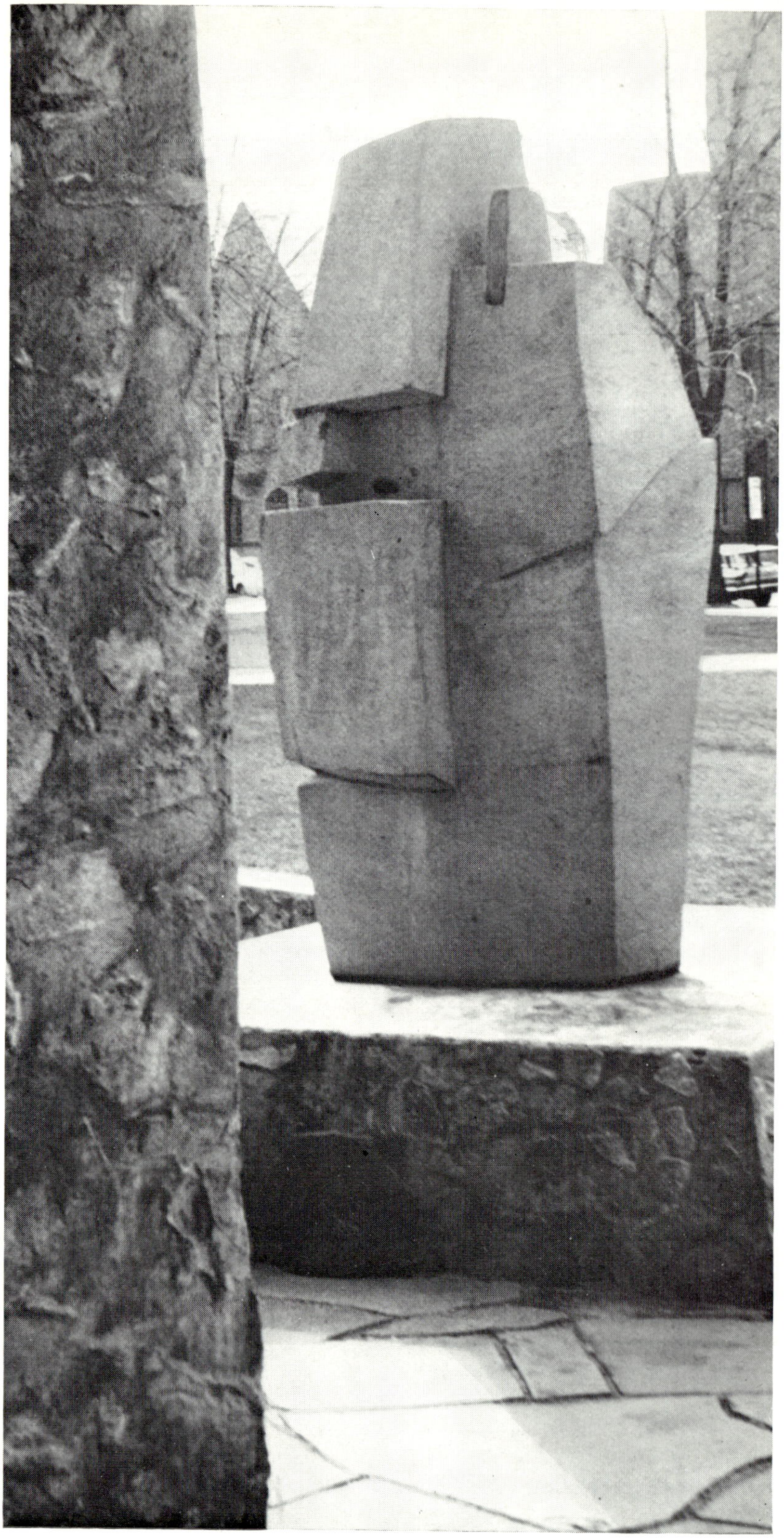

Plates 23 and 24. Two sculptures by Constantino Nivola, made *in situ* for the residential building at Yale University, New Haven, Connecticutt (Approximately 8ft. high). These sculptures are two of a great number designed by the sculptor to fit in with the architecture of Saarinen. They illustrate a very successful interaction.

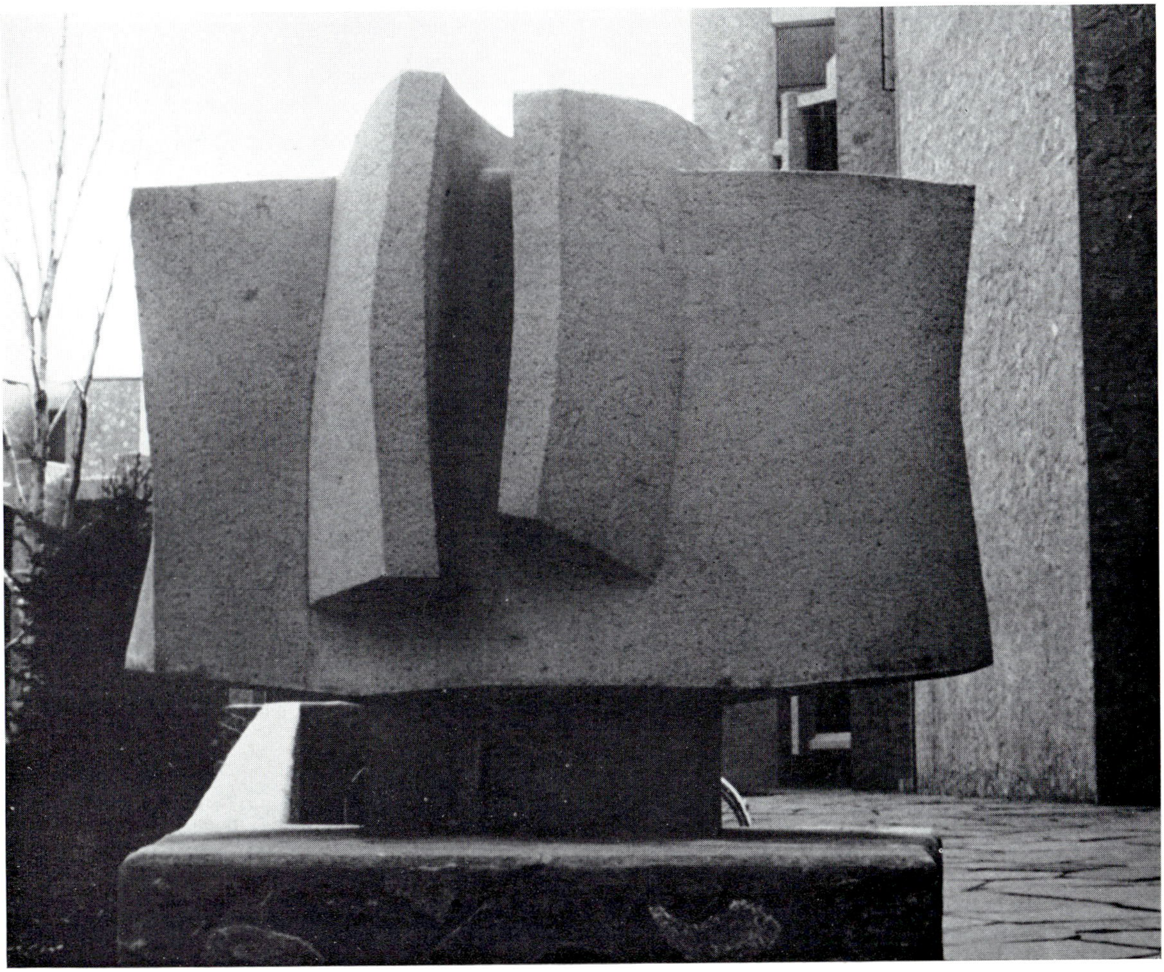

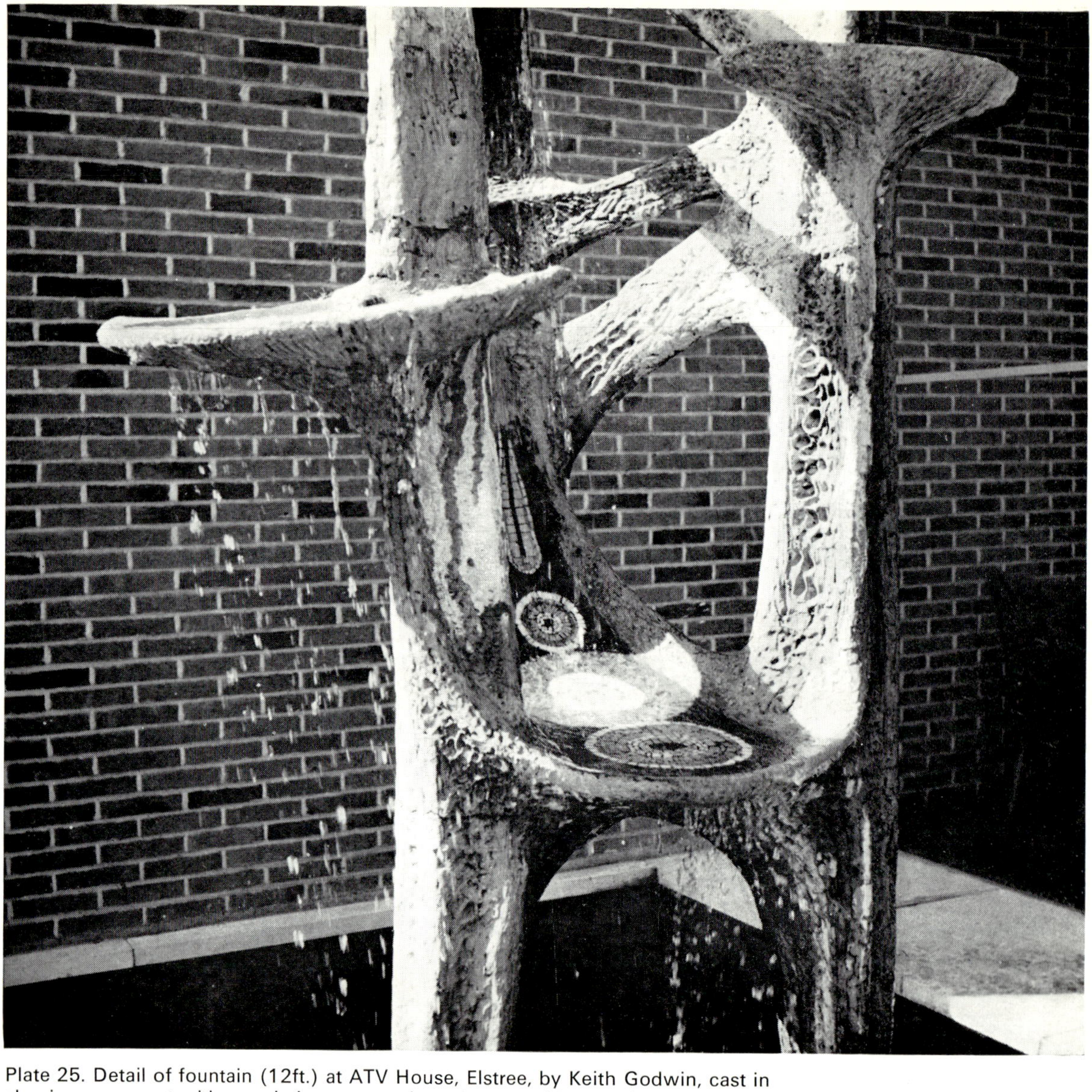

Plate 25. Detail of fountain (12ft.) at ATV House, Elstree, by Keith Godwin, cast in aluminous cement with mosaic inserts, cast in and exposed by rubbing down the casting.

FIG. 9. A cambed bit made to fit an electric drill to make a simple vibrator in the studio.

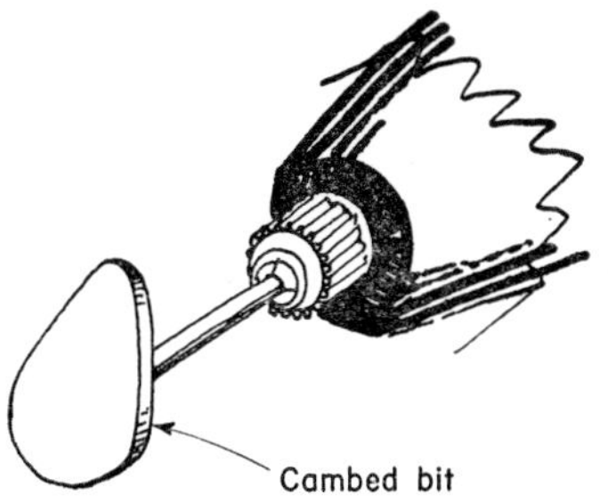

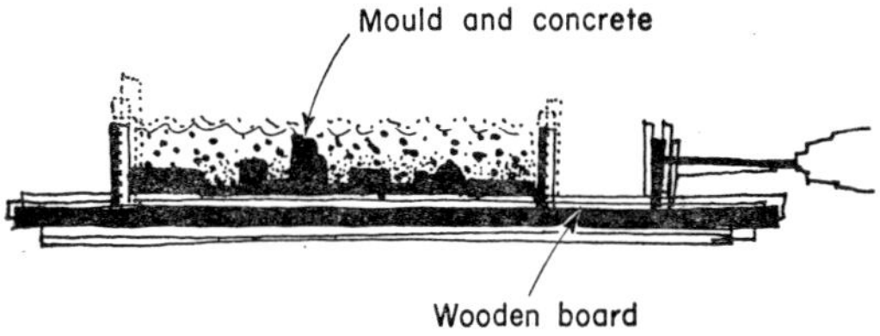

Large aggregate casting

Edward Folkard must be given credit for helping to establish concrete in the repertoire of the sculptor's material. Using mainly high alumina cements he has taught a great number of young sculptors the techniques of using large aggregates. I shall describe this technique in general terms, but shall not give descriptions for filling particular forms. The individual can easily adapt the skill described to suit his or her particular need.

The mould is prepared, treated with a release agent, and soaked.

The concrete mix is determined and a 'dry mix' sufficient to do the work in hand, is mixed to be even in colour throughout. The work in hand may be the complete filling of both 'caps' and 'main mould', or, according to the size of the work, perhaps only part.

Next, a paste of neat cement and water is made. This has come to be known as 'goo'. It needs to be the consistency of thick cream. As it is applied with a brush, it must not be too stiff. A paste, or 'goo', of cement and fine aggregate can be used instead of neat cement and water (see Fig. 10).

Water must now be added to the 'dry mix' to make the 'wet mix'. Small quantities can be made from the 'dry mix', or the entire mix can be made wet. This will be governed by the size and area of the work, or section of work, to be filled.

Commencing with the caps, a substantial layer of 'goo' is applied to the surface of the soaked mould, being brushed well into texture and detail. This should be evenly applied to cover any high points on the mould surface (these, of course, will be hollows on the positive form).

On to this neat cement the larger aggregate mix is placed, and compacted by firmly 'tamping'. The required cast thickness is arrived at by careful compacting and building.

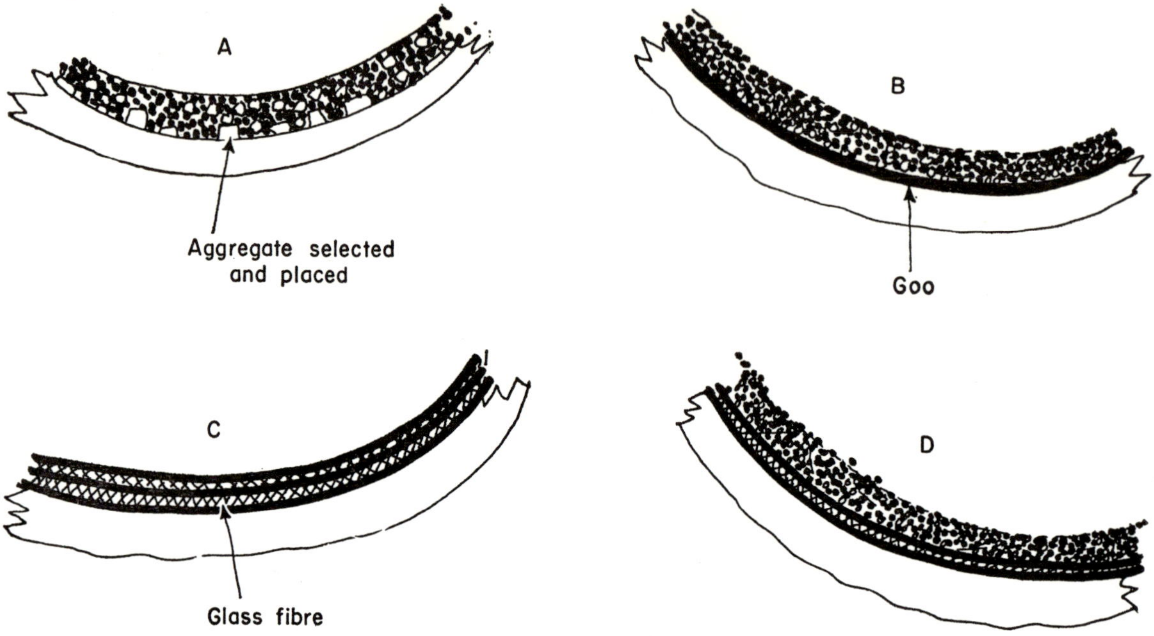

FIG. 10. FILLINGS. (A) Illustrates a coarse aggregate mix placed directly to the mould surface. Some selected large aggregate can be placed as the casting is made, to be exposed later.
(B) Shows a coarse aggregate placed over a 'goo' or layer of fine aggregate mix to ensure good reproduction and no honeycombing.
(C) Illustrates the laminations of fine aggregate mix and glass fibre. The fibre must be well impregnated with concrete.
(D) Is a mixture of the techniques shown in (B) and (C).

Mild-steel reinforcing should be built into this concrete thickness, and be well covered.

The 'tamping' to compact the concrete will make a dull ring when it is sufficient. A wooden mallet is useful for this. Finer wooden implements can be made to 'tamp' in small restricted forms.

Remove any excess moisture that makes its way up from the mould surface. If it remains it may weaken the concrete.

Complete all the caps, up to the seam edges.

These edges are of the utmost importance, throughout the moulding-casting process. They must be formed in such a way as to permit the closure of the mould, by fitting the caps to the main mould, also permitting proper filling at the seams, to join and fill the sections of casting. The plaster faces must be washed clean of any obstructions (see Fig. 11).

The main mould is treated in the same way:

1 'Goo' is applied evenly over the surface.
2 The large aggregate mix is built up on this, while the 'goo' is wet, and compacted to make the concrete thickness.
3 Mild steel reinforcing is made and placed. It must be built integrally in the concrete thickness. The main mould will take most of the reinforcing, because it contains the bulk of the form.

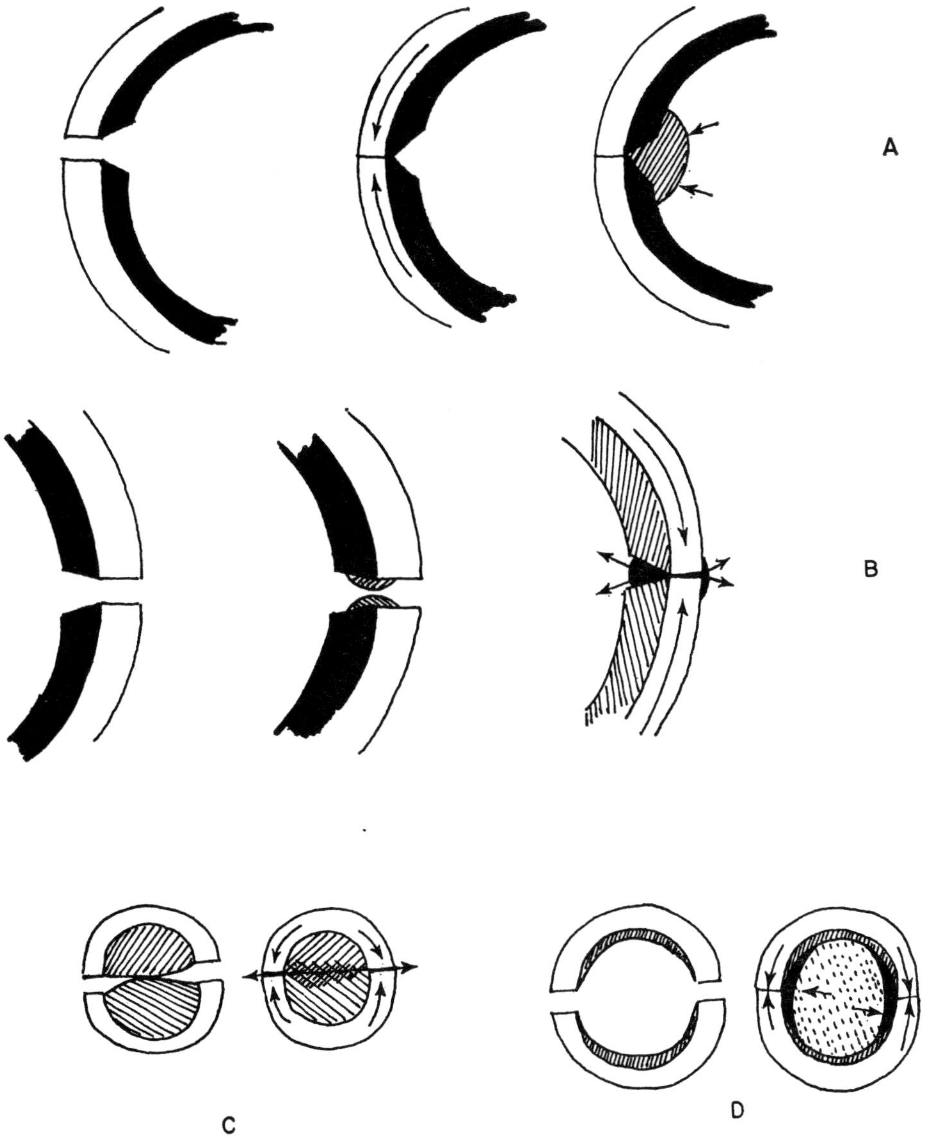

FIG. 11. SEAMS. (A) Shows the angle of the filling at the seam edge to be filled from the inside by packing.
(B) Illustrates the angle of filling at the seam edge for squeezing the sections.
(C) Shows the overfilling of small mould sections to be squeezed and cast solid.
(D) Shows an alternative to (A) when a more even cast thickness is required.
Combinations of these methods of filling will be used according to the nature of the mould and form. The angle at the seam edge must be planned to give maximum adhesion and the strongest possible casting.

4 The seam edges are then formed and the plaster seam faces cleaned off.

Overhanging surfaces can be filled in stages, filling section by section, rolling or turning the mould on its cradle when previous sections have set (see Fig. 12).

The caps and main mould, filled and completed, must be covered with wet cloths and plastic sheet, to encourage proper setting and hardening (see page 22).

When the sections have hardened they are ready to be fixed together to complete the casting process. This fixing of the caps to the main mould is effected in one or a combination of the following techniques:

Pouring Packing Squeezing

Pouring

Secure the caps to the main mould, making sure of proper register and secure them by tying, using plaster of Paris and scrim, or by clamping (see Fig. 13). A cavity is now formed within the inner surfaces of concrete filling, which should be wet. Concrete can be poured into this cavity to fill it, and the concrete 'tamped' to consolidate it. This technique has been used, for instance, for filling the legs of a standing figure, with a mild-steel reinforcing in the centre of the forms.

FIG. 12. To fill an overhang the lower surface is filled first as in (A). When this has set the mould is rolled to fill what was the overhanging surface (B).

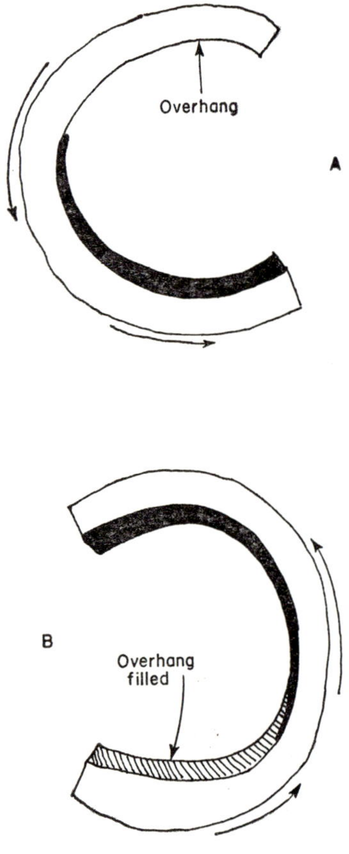

Packing

To pack, first secure the caps to the main mould (see Fig. 13), making sure of proper register. Secure one cap at a time. When secure the presented opening permits filling the seams by packing, which is done by applying firstly the 'goo' and secondly a large aggregate filling. This joins the sections of casting and fills the seam (see Fig. 11).

Squeezing

The squeezed filling of the seams is injected as the caps are placed and secured to the main mould. The angles of the seam faces must be planned to leave as small a space as possible to be filled. Up to these edges, on both the main mould and caps, a layer of 'goo' is built up to overlap concrete and plaster. The cap is placed on to the main mould, and the neat cement is squeezed to fill the seam. It squeezes out all round the mould and into the casting to fill the seam.

The most satisfactory method of fixing and filling caps and seams, is a combination of squeezing and packing, but this can only be done if the inner surfaces of the filling are accessible. These processes apply to all casting media.

Fine aggregate casting

Fine aggregates of dust-sieve size, mixed in the ratio 1 :1, can be wet mixed to make a paste that can be easily built up on the surface of the mould. This paste makes the casting process fairly simple. There is no need for haste because the relatively slow setting speed of concrete does not call for speed, and application of the filler to the mould surface can be evenly paced. The build-up of the cast thickness can be carried out methodically too. Glass fibre is always used with these mixes to give tensile strength.

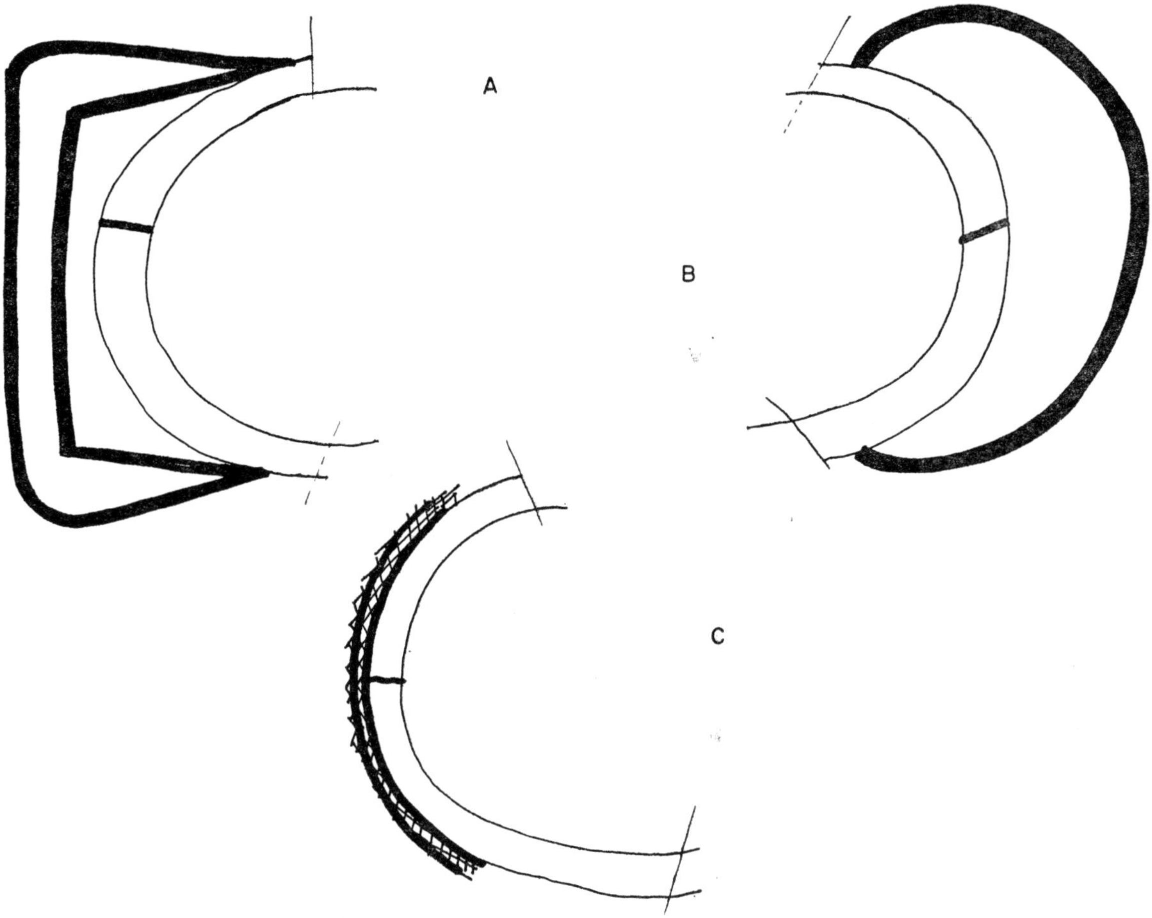

FIG. 13. Method of securing mould sections for filling.
(A) Carpenter's dog-clamp by its very shape gradually squeezes the mould tight as it is pushed on.
(B) A simple mild-steel clamp sprung on to the sections holds them tight.
(C) Scrim and plaster provides a very strong fixing.

The general casting process using fine dust aggregates is given here, and particular castings can be worked out fairly simply from the following description:

The mould is prepared; the necessary release agent applied, and the mould soaked. It is wise to start with the caps, so that the filling in these is firm and ready to be placed to the main mould when filled.

Dry mix sufficient concrete to complete the work immediately in hand.

From this 'dry mix' make a small bowlful of 'wet mix' (goo). This needs to be a paste or thick cream, capable of being brushed on to a mould surface. Another bowl of 'wet mix', stiffer in consistency and of a larger volume should be made. These mixes can be made while the mould is soaking.

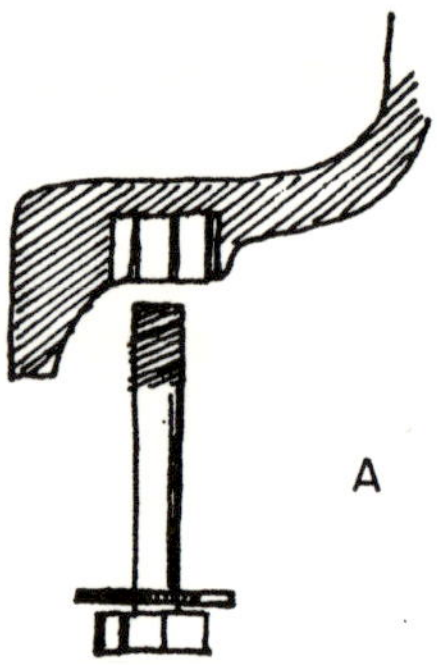

A

Chopped strand mat glass fibre is also prepared at this stage. The manufactured mat can be easily divided to produce two thinner layers. This should be done and shapes cut to fit the mould. The mat can also be torn to fit.

Apply a substantial layer of 'goo', $\frac{3}{16}$in.–$\frac{1}{4}$in. (0·6350 cm.) to the mould surface, beginning with the caps. This must be evenly applied to cover the surface, taking care to cover the high points of the mould (*low* points on the positive form). Do each cap separately, and do not try to cover too great an area at first. Use a brush for this first layer.

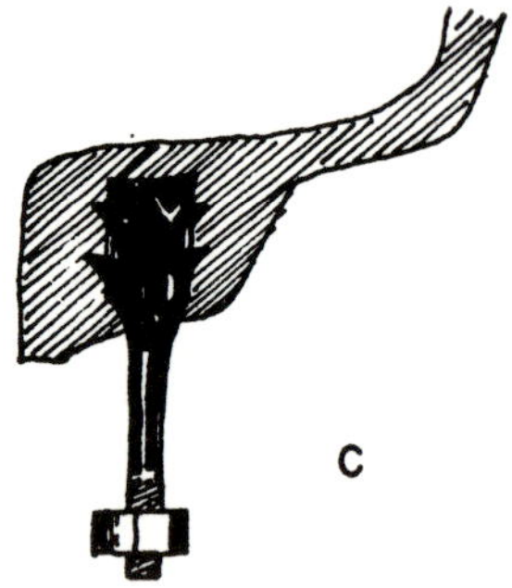

C

Into this thickness of concrete the first of the two layers of laminates of glass fibre is placed. By carefully placing and patting, the fibre is made to cover the concrete. It must not be pushed down on to the surface of the mould.

Take a brush and stipple the fibre with 'goo' to impregnate it.

Now, using the stiffer mix, build up a thickness of concrete evenly and carefully over the first layers. This needs to build the cast thickness up to a little more than $\frac{1}{4}$in., but it may be varied according to thickness of the first layer.

(Any mild-steel reinforcing can be shaped and placed at this stage, but most of this is contained in the main mould.)

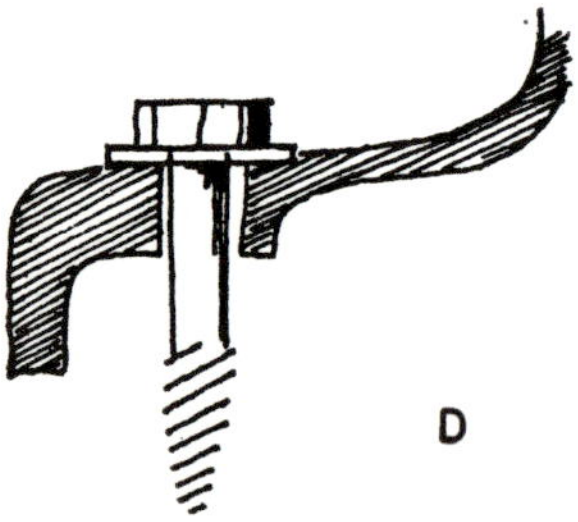

D

On to this second layer of concrete, another layer of fibre is placed, again by patting and stippling.

Then with the stiffer mix, build up the cast thickness. This build-up should be trowelled to a tight surface. In this way a double surface tension is achieved in the cast thickness, giving additional strength. Now shape the seam edges.

Any excess moisture that appears on the concrete as it is worked, should be mopped up, and should not be left to remain as it might well weaken the casting.

The caps and main mould should be treated as described, building a cast thickness in the following stages:

1 Prepare a 'goo' and a stiff concrete mix.
2 Paint a layer of 'goo' to the prepared mould, evenly.
3 Carefully place glass fibre mat, patting and stippling to impregnate.
4 Apply a second layer of concrete from the stiffer mix.
5 Apply the second layer of fibre.
6 Build up the cast thickness, including any mild-steel reinforcing in this build-up. Trowel to a tight surface finish.
7 Shape, trim, and clean the seam edges (see Fig. 2, and page 36).
8 Complete the casting by fixing caps to main mould.

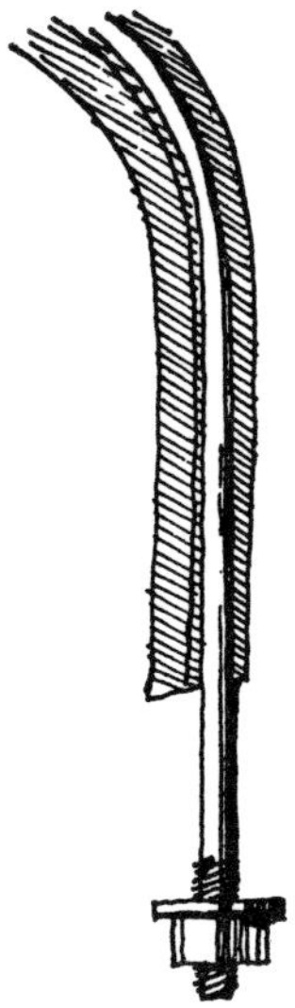

B

FIG. 14. Various fixings that can be used to secure concrete sculptures. (A) The nut is included in the casting, protecting the thread with the bolt in place well greased. (B) A long rod used in a slender form, or included in any cast thickness, threaded for a bolt at the exposed end for fixing. (C) Rag bolt, barbed to key to the concrete when cast into the sculpture. This can be grouted into a hole drilled in the plinth. (D) Coach screw and washer fixed through a clearance hole in the casting. (E) A wood screw fixed through a countersunk clearance hole. This can be filled when the job is secured. Both (D) and (E) are for fixing to wooden bases.

This final moment of completing the casting by fixing the caps, is the critical stage of the process. The seam edges must be clean, and designed according to the final processes. Glass fibre overlap, can be folded back to give a little extra strength to the seam edge. The method combining squeezing and packing (see page 32) is the most useful. There is no need to wait for the filling to harden. After the initial set, the caps can be fitted and filled without risk of the concrete dropping. Overhanging sections of the mould can be filled also with comparative ease.

When the mould-filling is completed, cover with wet cloths and plastic sheet to gain proper setting, hardening and cure (see page 22).

A comparison of the two casting techniques described will emphasize the contrast of methods. It may well be that the job planned demands a combination of the two. No two works of sculpture are exactly the same. Differences in image and form make various demands, and a good working knowledge of the casting processes available enlarges the scope and range of expression.

Both techniques can be used to make castings from waste-moulds, piece moulds of plaster of Paris or reinforced polyester resin, from the various kinds of flexible mould, and from built up moulds i.e., shuttering or moulds built *in situ* for direct working. But proper allowance must always be made for release and hydration.

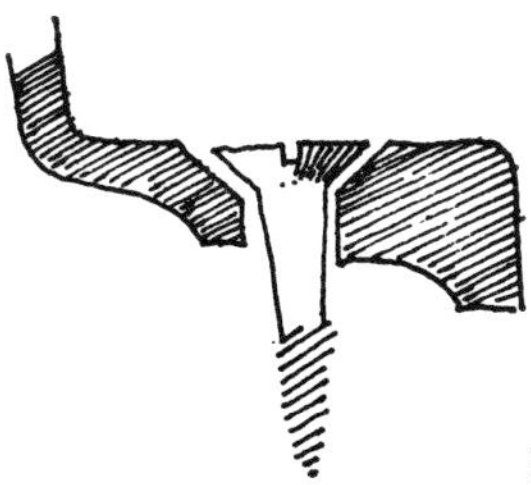

E

Solid castings

Solid castings are seldom made on a large scale, for the reasons already given. Very small castings, however, are often made solid for expediency. These are simply made by overfilling the sections of prepared mould, usually consisting of only two pieces, using the 'goo' and the stiffer mix with fibre included. The overfill is built up on the sections which are immediately squeezed together. Excess filling is forced out and the sections are secured together during setting, hardening and cure.

To fill a relief using either of the two filling techniques described, is a straightforward matter. Particular care must be taken to reinforce the casting well, because of the vulnerability of a comparatively large expanse. A small simple relief is often made solid.

Relief casting offers useful exercise for filling, presenting few complications other than a large area to be filled. Suitable fixing must be allowed for and possibly cast into the concrete thickness if it is to be placed *in situ* or mounted.

Means of fixing a sculpture are best cast into the work, and some suggestions are illustrated in Fig. 14.

4 After Casting

No matter how good the craftsman, no matter what the technique used, castings always need additional work to make good or finalize the surface and form. A casting is seldom removed from the mould requiring no further work. This factor often causes dismay to students, and to others unfamiliar with casting processes. Of course, perfection should be the aim but some flaw is inevitable, and near perfection can sometimes be achieved.

Perfection in itself can be dull. It is in the attempt to achieve it that the challenge lies.

Making good

Knowing what to do when things do not go perfectly, is a major part of any craftsman's skill.

The least work that can be expected on a casting, is the removal of seam flash, and here the value of careful filling and attention to the seam faces during filling can be seen. The flash is the excess filling forced between the plaster of Paris seam faces. This should be no thicker than the brass shim or clay wash used originally to make the mould divisions. That is perfection. Usually the flash is a little thicker. It is removed by breaking off large flash, then grinding down with carborundum till it is level with adjacent surfaces. On the fine aggregate castings the flash can be filed or rasped off, using marble rifflers or sharp coarse files. Wet and dry emery paper can be used to finish off any concrete. Power tools with grinding wheels are useful, electric drills, angle grinders. and flexible shaft tools can ease the task of cleaning off and grinding the excess concrete.

If the seam filling is not good, there will be places without filling, presenting unsightly holes. Such holes and any others that might appear on the casting can be filled, using the same concrete mix – neat cement 'goo' or cement and a fine aggregate 'goo'. The casting must be wet and any making-good kept wet to set, harden, and cure properly. This is the best method, ensuring consistency throughout the work.

If, when an exposed aggregate is used, the surface fines when removed reveal any honeycombing, the whole casting must be made wet before this surface is filled. The filling is done using the same method that stone carvers use to fill a coarse stone or marble surface. Neat cement, or cement and fine aggregate is made into a paste. This paste is then rubbed into the honeycombed surface, to overfill it. It is then left to set and harden, when it is rubbed down to make a uniform surface. Wet and dry emery paper and fine carborundum stone is used.

Any making-good on the surface must be well keyed, and the surface to be treated roughened to enable the fresh filler to bond well. It must also be wet.

Plastic adhesives can be used to make satisfactory fillers. There are various PVA (polyvinyl acetate) adhesives on the market, which can be used with the necessary cement to make a filler, and are useful for making good any defects on the castings.

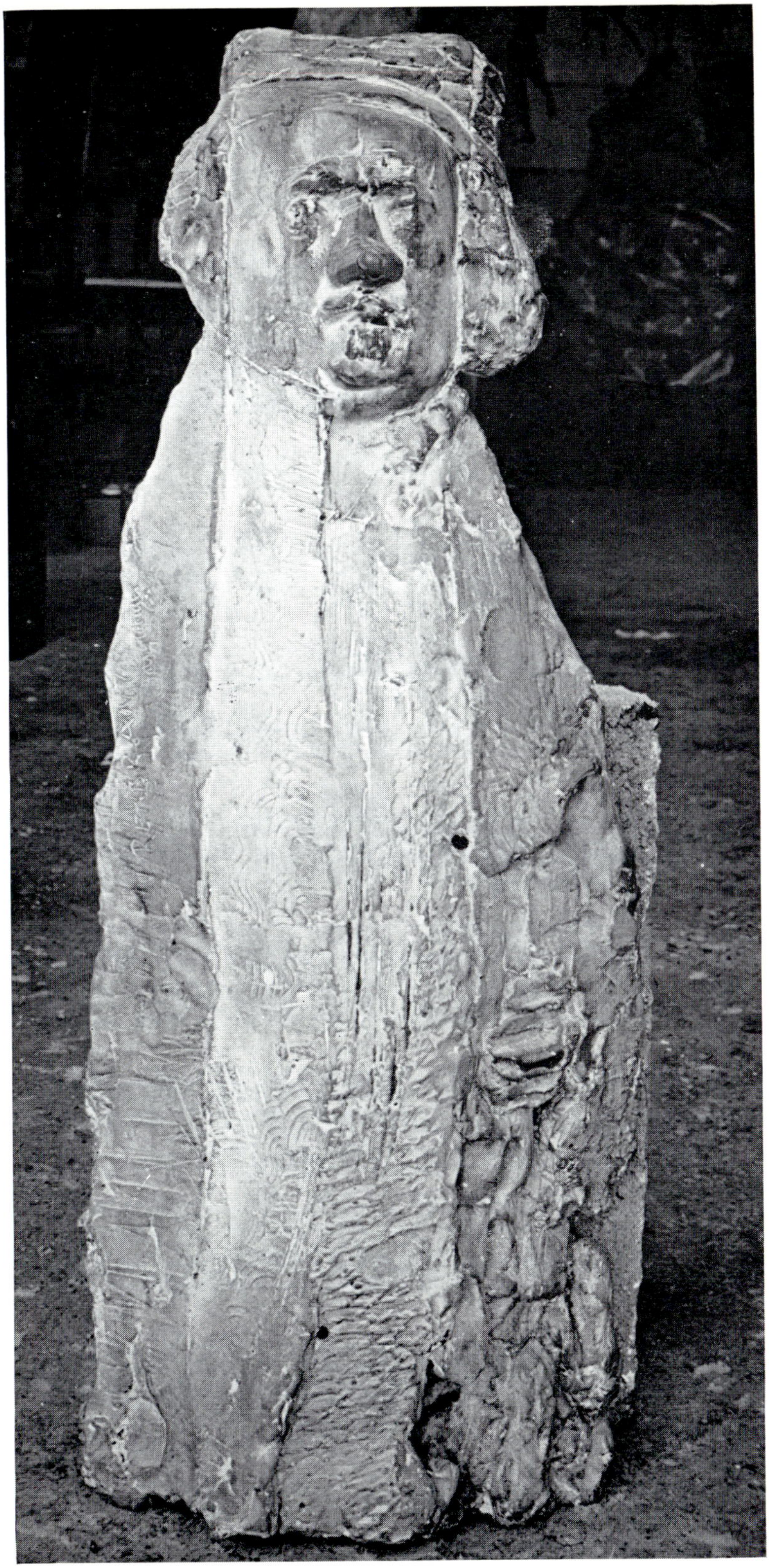

Plate 26. 'Van Rijn', aluminous cement, (Life size) by John W. Mills.

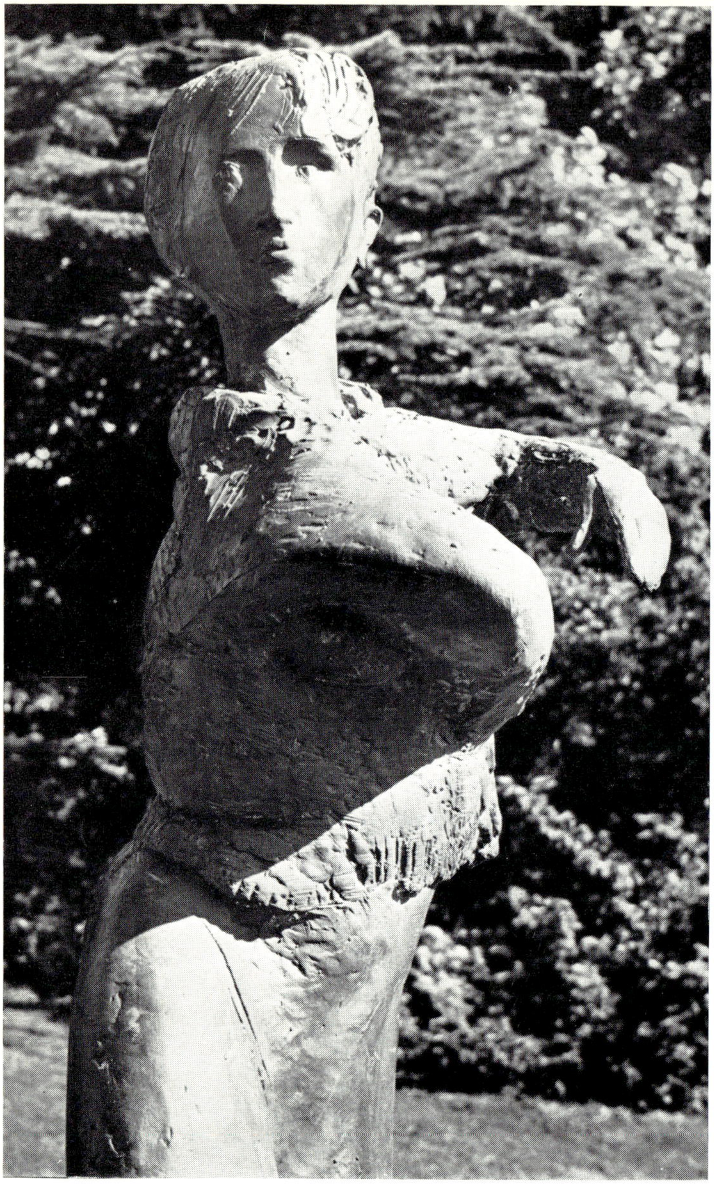

Plate 27. (Left) Detail of 'Dancer Turning', aluminous cement (5ft. 6in.). In the collection of Queenswood School.

Plate 28. (Below) Maquette (1ft. 6in.) for 'Orson Welles', aluminous cement. By John W. Mills.

Plate 29. (Facing page) 'Orson Welles', aluminous cement. (Height 3ft. 6in.) John W. Mills, outside the studio receiving a patina and final surface.

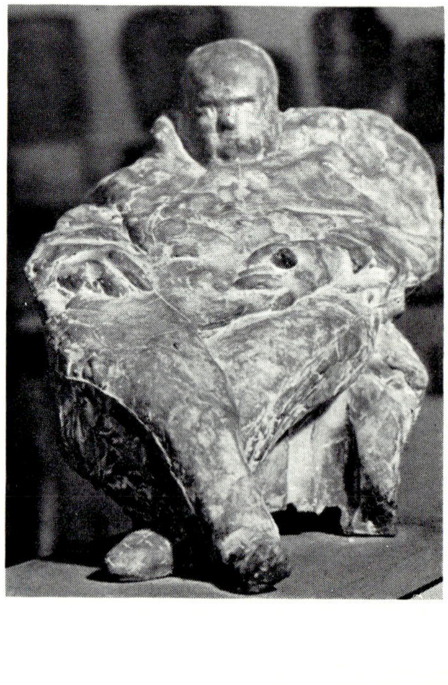

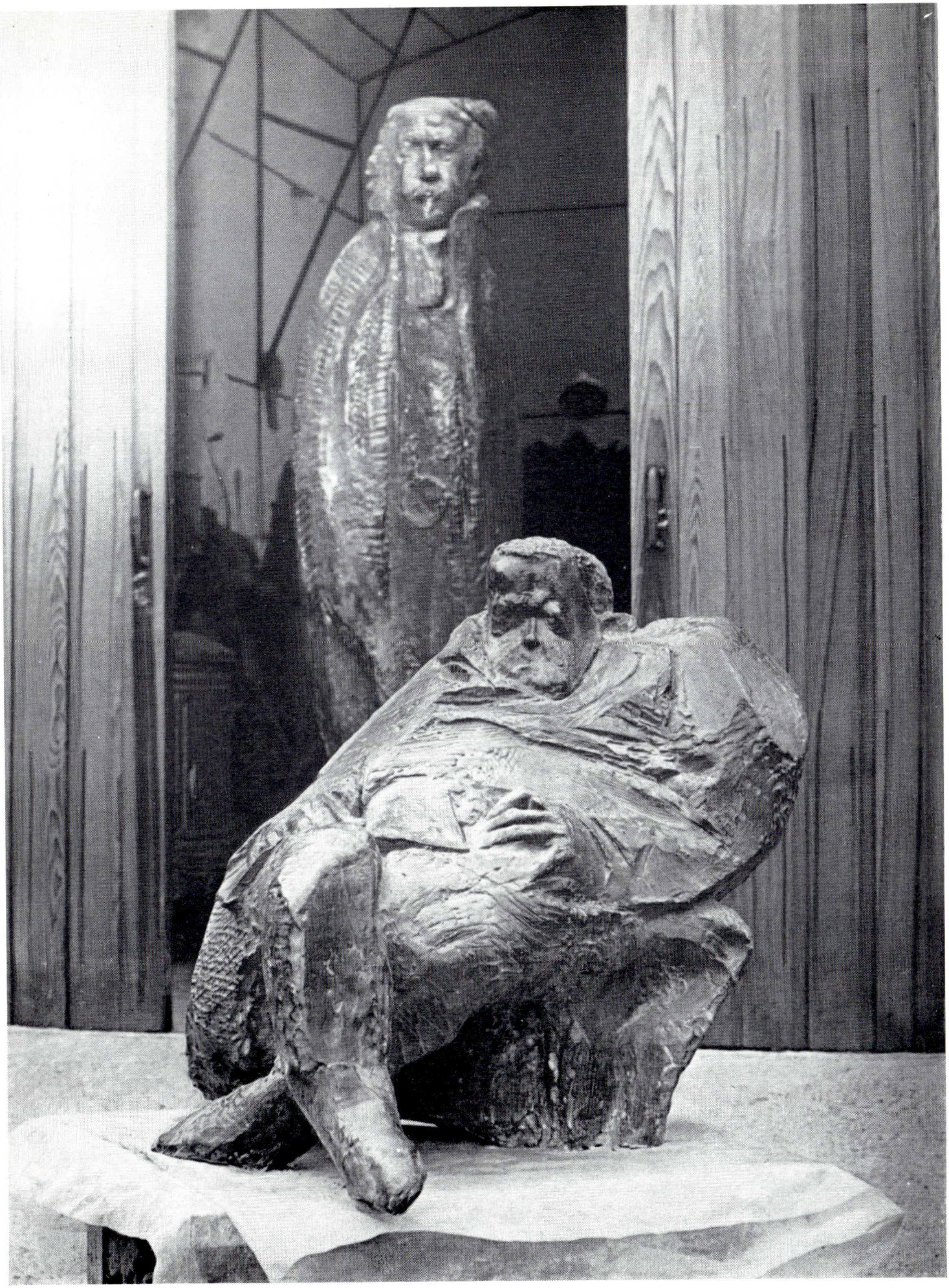

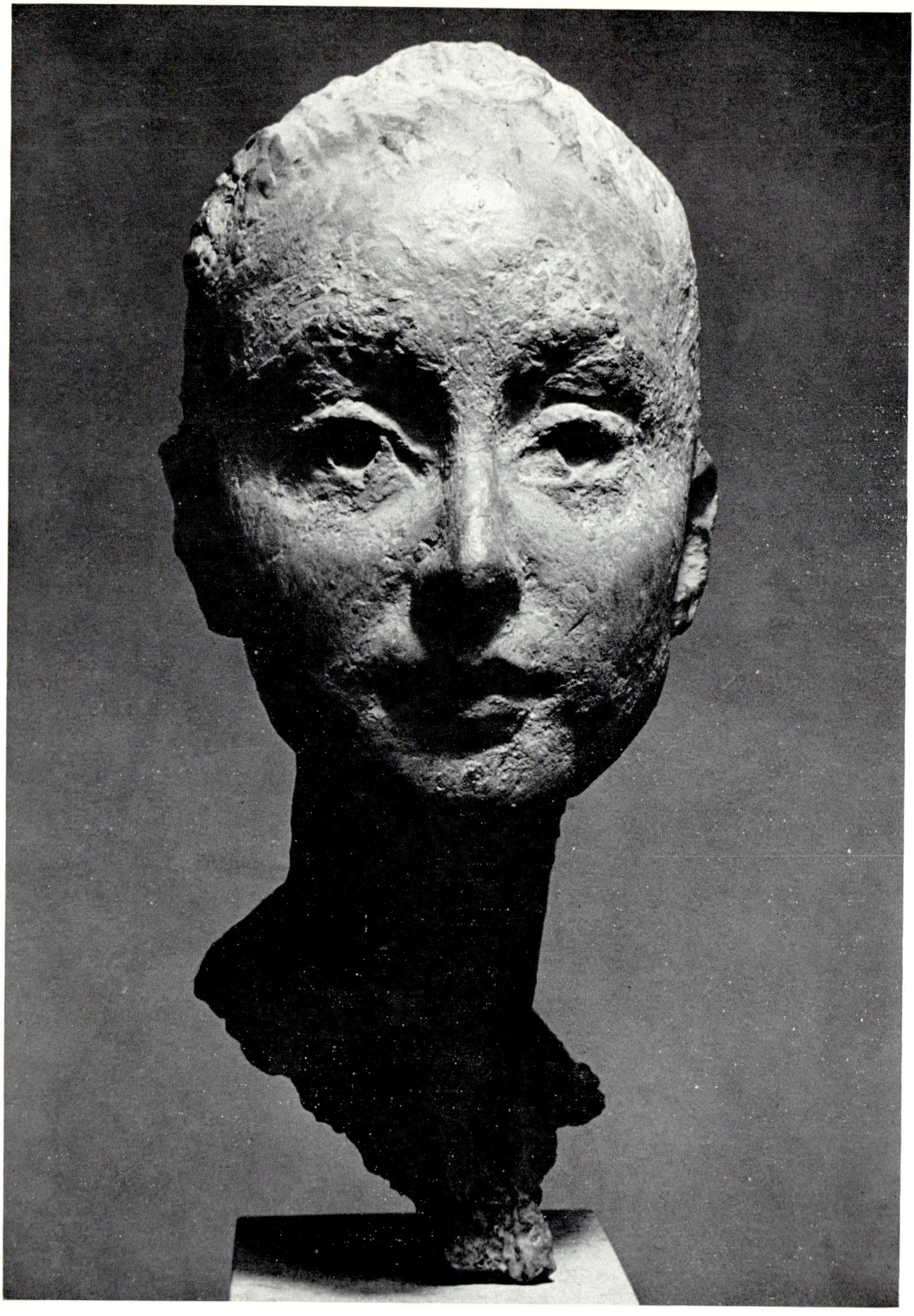

Plate 30. 'Portrait of a Girl', aluminous cement, by Sydney Harpley.

It is wise to ask for details of any PVA adhesive from the building suppliers, who will know what is available in particular areas, or countries. In the United Kingdom there are two proprietary brands 'Uni-bond' and 'Polybond'. Some paint manufacturers will supply a basic PVA which may prove cheaper.

A dilute solution of 10 water : 1 PVA is first used to prime the surface needing repair. This should be roughened and keyed first, but it need not be wet. Then by mixing the filler to stiff clay or putty consistency, the repair can be modelled. A steel spatula is the best tool to use, plus the fingers. Dipping fingers and spatula into water first, additions can be made to the surface. By this method cement and concrete can also be made to bond firmly on to metal.

Large holes through castings sometimes occur and always present a problem. A simple way to deal with this defect is as follows (see Fig. 15) :

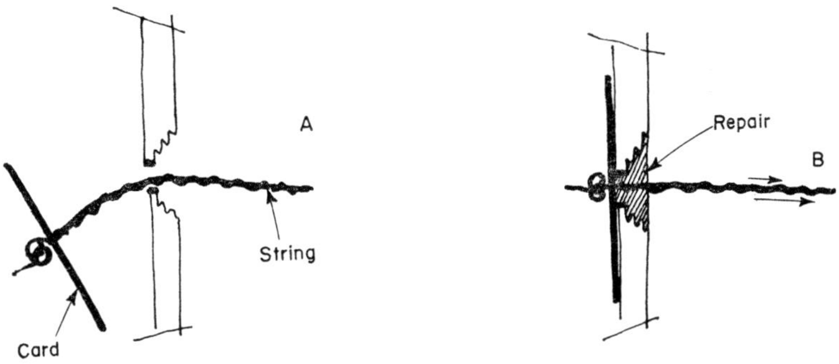

FIG. 15. Repairing a hole through the cast thickness. The cardboard backing is threaded on a string (A) and pushed through the hole. When the string is pulled tight the cardboard presses against the inside surface (B) providing a backing for the repair.

1 Cut a piece of thin cardboard slightly larger than the hole and to fit over it.
2 Thread a string through the middle of this card, with a knot to prevent it pulling through.
3 Force the card through the hole and pull the string hard to bring the card close against the inner cast surface.
4 A firm backing is now established, preventing any filling from falling through into the cast. The repair can now be made with suitable filler.
5 When the repair has hardened, cut away the string.

Epoxy and polyester resins, with suitable fillers, are often used to make good defects on a cast surface. Such resins are used by stone masons to fill and to fix stone, to repair or to add to stone. The fillers for such resins must be inert, but almost any colour can be produced. Grindings from the actual casting, for instance, can be used. The casting must be dry if these resins are used because water inhibits their hardening.

There are various admixtures for concrete on the market that have been devised to ensure properties not inherent in the material. They are made for numerous reasons, and should be investigated.

Patination

The 'patina' or final surface texture and colour is the ultimate consideration of the work before presentation. The range is of course infinite. Integral qualities of colour and texture can be achieved according to the concrete mix, and casting method. In addition there are the possibilities of painting a surface to acquire a particular colour or effect. The range is wide.

Naturally, the patina that is arrived at by exploitation of the material without painted addition is the most lasting and usually the most suitable. Certainly it is the most likely to stand up to open air conditions. The fact that concrete is, beyond doubt, established and familiar throughout the world, manifesting its inherent qualities in its own right, makes it rather ridiculous to describe applied patinas that simulate metals or stones. There may sometimes be valid reasons for requiring these exotic finishes, and for such purposes the techniques for simulating metals and marquettes etc., are described adequately in my book *The Technique of Sculpture* – techniques which can be used on any solid.

Colour and texture arising from inherent qualities and techniques of exposed aggregates have been described and Mr Wilson's book on *Exposed Concrete Finishes* is recommended. The colour of the basic cements, grey for Portland cements, nearly black for high alumina cements, can be supplemented by white Portland and white alumina cements. Both are available and can be mixed with their own kind, but they are expensive.

Plaster-bloom picked up from the mould, is a quality peculiar to cast concrete, and is worth exploiting. The bloom can be tinted and coloured effectively, or removed in specific areas. If a great deal of repair is necessary, however, it will spoil the chance of using the bloom.

A solution of whiting and milk, stippled on. may help to simulate the bloom in small areas.

A technique developed by the author personally is that of applying a dilute vinyl glaze to the cleaned casting. This seals the surface, but does not obscure the colour. When this glaze, which is fairly flexible and not liable to crack, is hard, oil or emulsion paint can be applied. By developing a technique of stippling the paint on and rubbing a little off, three qualities are exploited : the applied colour or colours, the dull glaze (vina glaze) and the concrete colour, bloom, and texture which can be seen through the glaze.

The final colour and texture of a sculpture is such a personal matter that it is pointless to lay down hard and fast rules. Experiment is the important thing. Sensible experiment to arrive at a personal style and technique is as important to patination, as it is to the modelling of the sculptured image. Experiment yourself – that is the best advice for the sculptor.

Direct working in concrete

To build a sculpture direct in concrete requires a technique that is difficult and arduous. The control of a gradual build-up to the desired volume and form, must be very strict. The descriptions given of actual casting methods indicate the workability of the material. It can be built up from any of the fine aggregates, and some of the smaller coarse aggregate (sand) mixes, over a well designed mild-steel armature. Metal trowels and steel spatulas are the best modelling tools.

The armature for direct working is best made by welding. In this way a strong armature can be designed and made with a degree of freedom. Expanded metal in various gauges is necessary to provide suitable binding of concrete to the armature, and so prevent the concrete dropping before it sets. The armature must be covered with concrete. Certain forms can be cast *in situ* during the build-up by constructing a suitable mould (shuttering, formwork).

The traditional stone carving tools and techniques can be used to finalize the form, exploiting the additive and subtractive processes of sculpture. Points to watch when building direct are:

- Suitably strong armature.
- Sensible concrete mix.
- Planned build-up to the desired volume and form.
- Proper adhesion between applications of concrete by putting a paste of neat cement between.
- Good control of moisture to ensure setting, hardening and curing. The work must be wetted and keyed before each new application of concrete.
- Complete coverage of the armature.
- Control of the carving tools and abrasives to finalize the surface, texture etc.

The forms must be strong enough to carve. Experiment is the order of the day. Wise experiment can only be beneficial, and should be made before embarking on any important work.

Glossary

Glossary

ASSEMBLAGE Sculptural form made by assembling various shapes and materials.

ARMATURE The internal support to a form.

AGGREGATE Inert granular material mixed with cement and water to make concrete.

CAST The form produced by filling a mould. The positive form.

CASTING The process of making a cast.

CEMENT The binding agent used to make concrete.

CONCRETE A solid made by mixing cement, aggregate and water.

CAPS The small removable sections of a waste mould, sometimes known as traps.

CLAY WALL A technique of moulding using clay strips to divide the mould sections.

DRY MIX The mixture of cement and aggregate before adding water.

FLASH The cast line excess at the seam on the positive form.

FLEXIBLE MOULD A mould made of gelatine, polyvinyl chloride, silicone or latex.

GOO The neat cement and water, or fine aggregate cement and water applied to the mould surface to reproduce detail.

GROG Ground ceramic sometimes used as an aggregate.

IN SITU On site. Constructed in position.

IRONS The term sometimes given to mild steel reinforcing used in a cast.

KEY The device for ensuring good register between mould sections.

LAMINATION A thickness built up in layers.

HYDRATION Combining chemically with water.

MOULD The negative form, from which the cast is made.

MORTAR A mixture of cement and sand.

MAIN MOULD That part of the mould containing the greater portion of the form.

PACK Placing the filling on to the mould surface.

PATINA The final surface texture and colour.

PORTLAND CEMENT Cement that resembles Portland stone.

PLASTER OF PARIS A fine white powder that sets hard when mixed with water 50:50.

PVA Polyvinyl acetate.

PVC Polyvinyl chloride.

RELEASE AGENT Material applied to a mould surface to prevent sticking, to cause surfaces to release. Sometimes known as parting agent.

RAG BOLT A textured bolt for fixing items to walls.

RELIEF A design made to relieve the flat surface. It can project or be incised.

SCRIM Coarse woven fabric usually jute, used to reinforce plaster of Paris.

SEAMS The dividing lines of a mould.

SHELLAC Varnish made from the secretion of an insect, the lac.

SHUTTERING A mould usually of wood constructed in situ. Sometimes known as formwork.

SOFT SOAP A semi-liquid soap.

SHIM Metal foil, 5,000th gauge, used to make divisions in the mould.

SPATULA A simple metal tool ideal for use with plaster and cement. Sometimes known as a plasterer's small tool.

TAMPING The act of packing and consolidating a concrete mix.

WASTE MOULD A plaster of Paris mould which is chipped away from the cast.

WET MIX The mixture of cement and aggregate after adding water.